World Internet Development Report 2019

Chinese Academy of Cyberspace Studies

World Internet Development Report 2019

Blue Book for World Internet Conference
Translated by CCTB Translation Service

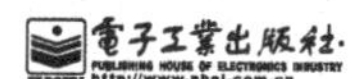

Chinese Academy of Cyberspace Studies
Beijing, China

ISBN 978-981-33-6937-5 ISBN 978-981-33-6938-2 (eBook)
https://doi.org/10.1007/978-981-33-6938-2

Distribution rights for print copies in China mainland: Publishing House of Electronics Industry Co., Ltd.

This Springer imprint is published by the registered company Springer Nature Singapore Pte Ltd.
The registered company address is: 152 Beach Road, #21-01/04 Gateway East, Singapore 189721, Singapore

Preface

The year 2019 marks the 50th anniversary of Internet. We earnestly compiled *World Internet Development Report 2019* (hereinafter referred to as the "Report"). As the third annual report on world Internet development since 2017, the Report adheres to the idea of China's Internet Governance and relies on the practice of Internet development in various countries. It analyzes, summarizes, and evaluates world Internet development in 2019, mainly in seven fields, namely information infrastructure, network information technology, digital economy, digital government, Internet media, cybersecurity, and international governance. It aims to comprehensively present the situation of Internet development in 2019 and to provide new ideas and intellectual support for world Internet development. The Report has three characteristics as below.

(1) It reviews the history of Internet and forecasts its future. The Report starts with the part "Fifty Years of World Internet Development," which briefly reviews Internet development in the past 50 years, describes its profound impact on economy and society, and analyzes the opportunities and challenges it has produced. It also urges international community to take advantage of developmental law, promote technological innovation, and strengthen international governance so that Internet development can better benefit the world, human, and the future.

(2) It makes scientific proof and systematic evaluation. The Report goes on to evaluate world Internet development and adopts the indicator and index systems consistent with those in 2017 and 2018. In order to further improve the rationality of the frameworks and the accuracy of data sources, it optimizes some indicators according to reality. Meanwhile, it selects 48 countries for comprehensive analysis to enlarge the coverage and make more scientific evaluation, covering major economies in five continents and representative countries in Internet development, so as to reflect the overall situation of world Internet development more comprehensively, objectively, and accurately.

(3) It relies on practice and makes in-depth interpretation. At present, major countries in the world generally regard Internet as national strategic focus and priority developmental direction and speed up the pace of construction and development. They actively explore ways in improving information infrastructure, promoting innovation and creation, strengthening cybersecurity, and enhancing international governance. The Report collects and summarizes the practice and effect of Internet development in major countries in 2019. And, it analyzes and interprets them for references for world Internet development.

The Report is an important achievement of innovation in idea, theory, strategy, policy, and practice made by Chinese academic circles for world Internet development and governance. In the future, we will keep paying attention to the trend and advance of world Internet development and putting forward our analyses and views to contribute Chinese wisdom and strength to the construction of a community of shared future in cyberspace.

September 2019

Chinese Academy of Cyberspace Studies (CACS)
Beijing, China

Fifty Years of World Internet Development

Internet boasts one of the greatest inventions of human in the twentieth century. Like fire used in primitive age, iron used in agricultural age, and steam engine and electricity used in industrial age, Internet has resulted in unparalleled profound changes to human society and ushered in an information age full of vigor and vitality for human, since its birth in 1969. In the past 50 years, Internet has increasingly penetrated into all fields, such as politics, economy, society, culture, and military affairs, speeded up the flowing and sharing of labor, capital, energy, and other elements, promoted new substantive leap of social productivity, changed the ways of production and life of human profoundly, affected world's political and economic patterns greatly, and improved human's ability in understanding and transforming the world. In terms of development speed, popularity scope, and influence degree, Internet enjoys unmatchable advantage over other scientific and technological achievements, leading and opening a new age for human history.

Along with fifty years of Internet development come the fast development of information technologies and the rapid popularization of network applications. It is a prominent fact that Internet has played a significant and valuable role in economic development and social and cultural advance, and human's probing and understanding of Internet has deepened over time. In the 1960s, during the Cold War, in order to take the lead in military technology, the USA increased the investment into basic research and created ARPANET, the prototype of Internet. In the 1970s, as an important scientific and technological innovation and a new information exchange tool, Internet spread gradually to institutions of higher education and government agencies in the USA as well as Europe. In the late 1980s, Internet extended from science and technology to the fields of economic society. In particular, in the early 1990s, the birth of the World Wide Web and the formation of global connection promoted the widespread and commercialization of Internet, with the rapid rise of network applications and network economy. From the late 1990s to the early twenty-first century, Internet grew rapidly and gained popularity throughout the world, and representative Internet corporations emerged all over the world. In recent decade, the mobile and intelligent development of Internet has become obvious. New-generation information technologies, such as artificial intelligence (AI), Internet of things (IoT), and big data, developed rapidly and intersected and combined with biology, energy, materials, and neuroscience. This inspired collective technological changes that were green, intelligent, and ubiquitous. New technologies, new industries, new applications, new business models and forms emerged successively, entering a new stage of intergenerational transition, full penetration, accelerated innovation, and Internet of everything (IoE). With deepened development of world multi-polarization, economic globalization, cultural diversity, and social informatization, Internet is now turning to a "new territory" of national sovereignty, new space for production and life, new channel for information dissemination, new platform for cultural prosperity, new engine for economic development, new carrier for social governance, and new link for international cooperation. It has fully integrated into and profoundly changed the progress of human society.

1. In the Fifty Years of World Internet Development, It Drove Economic Development and Stimulated Industrial Transformation.

Internet has created new demand and supply, accelerated the reconstruction of production, employment, distribution, consumption, and other aspects, promoted great changes in the pattern of economic development and the style of human life, improved the degree of information sharing and the efficiency of resource allocation, and boosted the breakthrough of productivity and the profound adjustment of production relations. Notably, there occurs economic transformation and reconstruction in the world. Under such circumstance, digital industrialization and industrial digitalization accelerated the process. Data became strategic basic resources like oil and electricity, and digital economy, as a new engine of global economic development, became an economic activity that develops fastest, embodies the most innovative elements, and radiates most extensively. Since the 1990s, the USA has seized the opportunity of networking and digital development and realized economic prosperity. European countries and Japan followed it, promoted digital transformation vigorously, and made remarkable achievements. Developing countries have made full use of the advantage of backwardness in digital economy to narrow the gap with developed countries and promote rapid economic development. Countries all over the world actively have accelerated the deep integration of Internet and real economy, used new information technologies to transform and upgrade traditional industries, cultivated new growth drivers with informatization to achieve new advance, given full play to the amplification, superposition, and multiplication of digital economy, continuously deepened and extended cooperation on digital economy, and enhanced the rapid and healthy development of global digital economy. As Huawei's Global Connectivity Index (GCI) Report 2018 suggests, digital economy had realized a growth rate of 2.5 times of that in global GDP in the past 15 years and would reach $23 trillion by 2025.

2. In the Fifty Years of World Internet Development, It Led Scientific and Technological Innovation and Realized Integration Across Disciplines and Industries.

Internet plays a leading role in promoting innovation-driven development and guides the development direction of advanced science and technology in the world.

For 50 years, with the rapid rise of scientific and technological revolution and industrial transformation, new-generation information technologies represented by Internet developed rapidly. Basic technologies and cutting-edge technologies iterate and evolve rapidly. Advanced technologies, such as artificial intelligence (AI), blockchain, cloud computing, and quantum information, manifest the vitality of innovation. 5G-based big data, edge computing, and virtual reality developed quickly. These accelerated the digital, networking, and intelligent transformation at a deeper and wider level. A report of World Intellectual Property Organization (WIPO) shows that in the past 20 years, among the top 30 global patent applicant enterprises, 80% were Internet-related. Internet integrates with new energy technology, new material technology, and biological technology, with various innovation emerging. Particularly, intelligent processing technologies represented by brain-like computing speed up the development of deep learning, unmanned driving and robotics, which, in the future, will change human's production and life in a deeper level and a wider range. Having realized the indispensable role of key technological innovation in information field, all countries strive to seize the historical opportunity of a new round of scientific and technological revolution and to stay ahead in technological innovation.

3. In the Fifty Years of World Internet Development, It Subverted Mode of Dissemination and Promoted Cultural Prosperity.

Internet is immediate, open, and interactive. It becomes a main channel for information production and dissemination based on these characteristics. As the reality suggests, "everyone has a microphone; everyone is in We Media." Internet has changed the one-way and centralized mode of dissemination, resulting in profound changes in public opinion ecosystem, media landscape, and mode of dissemination. Omni-media keeps growing, with new forms of all-process, holographic, all-participant, and all-effect media. Information exists everywhere, covers everything, and reaches everyone. Particularly, the rapid development of mobile Internet contributes to the widespread of social media. According to Global Digital Reports 2019 released in February 2019, active users in global social media reached more than 3.5 billion, with a global penetration rate of 45%. Social media has strengthened its ability in evoking the public and affecting public opinion and facilitated the explosive growth and fission dissemination of information. In cyberspace, ideas, cultures, and information widely assemble and freely flow in digital form, which effectively promotes the exchange and learning among different cultures and civilizations and provides a vast space for protecting and displaying the diversity of human civilization. The digital production and networking dissemination of excellent cultural products help to popularize the excellent cultures of various countries and nationalities, to present the lively scene in which various cultures co-exist and communicate in the world.

4. In the Fifty Years of World Internet Development, It Promoted Inclusive Development and Improved People's Life.

The number of Internet users in the world has reached nearly 4.5 billion. More and more people take the advantage of Internet. Through Internet, they see the world, obtain information, exchange ideas, innovate and start a business, enrich life, and change destiny, constantly creating and seizing new opportunities for better life. Internet established information superhighway accessible to all. With ubiquitous network information access facilities, convenient "Internet +" travel information service, 24-h online mobile retailing, "one-stop" on-the-road travel experience, online digitalized learning environment, inclusive online medical service, and intelligent online pension care experience, Internet opened an new age of intelligent life for human, which greatly promoted the coordinate, open and sharable development among nations, regions, urban and rural areas, and people. Internet development contributes to the realization of UN's Transforming Our World: The 2030 Agenda for Sustainable Development. It constantly provides better Internet products and more convenient Internet services for eliminating poverty, promoting health, improving energy efficiency, and popularizing basic education. In addition, Internet creates more opportunities for fair development for people and builds helpful platform for people to fully participate in modern economic and social activities and better realize their own value.

5. In the Fifty Years of World Internet Development, It Improved Social Governance and Optimized Public Services.

Internet provided new platform for social governance and new channel for public participation in public affairs. Mobile Internet enlarged the coverage of public services. Cloud service provided more flexible way to build and run e-government system. Big data serves became an important tool to support government scientific decision-making and accurate management. These are strong supports for turning from one-way management to two-way interaction, from sample data to massive data in government governance, and for enhancing the modernized level of the national governance system and governance capacity. E-government has made great progress in the world since the US government put forward "e-government" plan in the 1990s. The USA promoted IT digital transformation in federal institutions and established modern digital government. China accelerated the construction of "Digital China," coordinated the development of e-government, and promoted opening and sharing of information resources. In other developed countries, the construction of digital government was also accelerated. Driven by Internet, governments made more scientific decisions, social governance became more accurate, public services became more efficient, and citizens got more open channels to participate in social governance, with their rights to know, participate, express, and supervise better guaranteed.

6. In the Fifty Years of World Internet Development, It Deepened International Cooperation and Promoted Peaceful Development.

The birth and development of Internet has produced great historical opportunities for the peaceful development of human society. In agricultural society, the strategic resource was land; in industrial society, energy resource; and in information society, data, and human wisdom. Land and energy resource are limited, yet data and human wisdom are unlimited. They can be shared and used for mutual benefit. Besides, human do not obtain data and wisdom by war or plunder, so it adds new elements and new impetus and provides new opportunities for the peaceful development of human. With Internet development, cyberspace and real space combine. The rapid evolution of informatization with globalization has enormously enhanced the global flows of information, capital, technology, talent, and other elements. International community has increasingly become a global village where all people are interdependent. Seeking peace, development, cooperation, and mutual benefit has increasingly become people's common hope and pursuit all over the world. Facing new opportunities and challenges brought about by Internet development, international community comes to realize that the problems on Internet need to be solved jointly and that opportunities must be shared. Exclusivist and monopoly mean no solution; sharing and co-governance mean future success. In the past 50 years, dialogues and negotiations on cyberspace got deeper. Among the countries related to the United Nations (UN), Group of 20 (G20), Asia-Pacific Economic Cooperation (APEC), and the "Belt and Road" Initiative (BRI), network cooperation deepened and the construction of global network infrastructure accelerated. The cooperation on extensive digital economy was conducted widely, the ability in cybersecurity was improved, and the idea of respecting cyber-sovereignty was deeply rooted among the people, signaling a fairer trend of the global cyberspace development and governance.

Internet is a "double-edged sword." While providing human society with development opportunities, it brings about various risks and challenges to political security, economic security, cultural security, social security, and national defense security, as well as citizen's legal rights and interests in cyberspace.

(1) Cyber-infiltration endangers the security of political power. Political stability is the foundation for the development of all countries and the happiness of people. Some countries use network to carry out large-scale network monitoring and network stealing, or even to interfere in other countries' internal affairs and stir up social upheaval, which seriously endangers national political security.
(2) Cyberattacks jeopardizes economic security. Critical information infrastructure in the fields of finance, energy, electricity, transportation, and communication serves as the central nerve system for the economic and social operation. On condition that it is destroyed or even paralyzed, national economic security and public interests will be endangered, and serious security accident will be caused, with disastrous consequences.
(3) Cyber-harmful information endangers cultural security. Illegal and unhealthy information, including Internet rumors, decadent culture, vulgar, mischievous, absurd, pornographic and violent messages, severely erode teenagers' physical and mental health, corrupt social values, mislead value orientation, and cause the frequent occurrence of Internet immorality and lack of credibility.
(4) Cyberterrorism and cyber-crimes endanger social security. Terrorist, separatist, and extremist use the network to incite, plan, organize, and execute violent and terrorist activities, which directly threatens the safety of life and property and social order. Computer viruses such as Trojans and ransomware spread in cyberspace. Cyber-fraud, hacking, infringement of intellectual property rights,

network theft, and misuse of personal information exist massively. Some organizations want only steal user's information, transaction data, location information, and business secrets, which badly harms the interests of the state, enterprises, and individuals and affects social harmony and stability.

(5) International competition in cyberspace endangers peace and security. Due to the rise of protectionism and unilateralism, international competition becomes increasingly fierce in plundering and controlling strategic resources in cyberspace, seizing the right to make rules and strategic favorable position, and seeking strategic priority. Some countries uphold "bully-stance," openly suppress or contain others, trample on international rules, strengthen cyber-deterrence strategy, constantly enhance the preparation for cyber-warfare and the construction of cyber-forces, and intensify arms race in cyberspace. As a result, militarization of the global cyberspace becomes even more violent and world peace faces new challenge.

Internet underwent 50 years of magnificent development. On a new historical starting point, opportunities and challenges co-exist, but the former is bigger than the latter. The curtain of information age has just opened. Full of human's vision and longing for better life, the train of Internet development is heading for the future. On the basis of mutual respect and trust, international community should make joint efforts to strengthen communication, expand consensus, deepen cooperation, seek common development and well-being, and face risks and challenges together. It also should work together to build a peaceful, safe, open, cooperative, and orderly cyberspace and a community of shared future in cyberspace, so that the opportunities and achievements of the prosperous Internet development can better benefit the world, human, and the future.

Overview

1 Overall Trends in World Internet Development in 2019

2019 is the 50th year of world Internet development. All countries consider cyberspace a critical field where they vie for leading position and build new advantage in future development and international competition. In general, world Internet presents a trend of seeking commonness in cooperation, advancing in setback, and developing in innovation. Infrastructure construction of 5G, IPv6, Internet of things (IoT), satellite Internet, and industrial Internet is developing steadily. They deeply integrate with emerging technologies such as artificial intelligence (AI), providing support to the development of digital economy and digital transformation and hastening economic and social development. Global digital economy displays vitality and accelerates the transformation of digital technology dividends. Cultural diversity in cyberspace enriches the spiritual life of human and promotes the integration and exchange of human civilization. The construction of cybersecurity protection capacity is commonly strengthened, and cybersecurity industries keep growing. It becomes an extensive consensus in international community to build a community of shared future in cyberspace, with a louder voice supporting cyber-sovereignty. Meanwhile, we must also notice that new risks to cybersecurity caused by cutting-edge technologies are getting prominent. Militarization of cyberspace becomes even increasingly intense, and traditional and new threats to cybersecurity intertwine. The vulnerability and uncertainty of cyberspace international rules appear continuously. This poses new challenges to word Internet development.

1.1 Steadily Progressing Digital Infrastructure Construction

In 2019, digital infrastructure construction represented by 5G and IPv6 steadily progressed. 5G is featured with enhanced mobile broadband (eMBB), massive machine-type communication (mMTC), and ultra-reliable low-latency communication (uRLLC). And IPv6 enjoys advantages of rich IP address resources, high security, strong application extensibility, and highly effective online forwarding. These provide technological support for the Internet of everything (IoE). Currently, telecom operators in many countries (regions) are actively promoting commercial 5G network deployment. By the end of June 2019, 5G test and experiment have been conducted by 280 telecom operators in 94 countries worldwide. And South Korea, the USA, Switzerland, Italy, the UK, the United Arab Emirates, Spain, and Kuwait have provided 5G commercial services. According to Qualcomm 5G Report, it is expected that by 2035, 5G will generate $12.3 trillion in global economic output, and that the output of China's 5G value chain will reach $984 billion, creating 9.5 million jobs and ranking first in the world.[1] Meanwhile, global network and service providers are also strengthening IPv6 deployment. According to Google's data, in 2018, more than 25% of all Internet-connected networks worldwide achieved IPv6 connection, and 24 countries (regions) provided more than 15% traffic through IPv6. Telecom operators in the USA, Japan, and India are promoting the use of IPv6 addresses, and China's major telecom operators have all achieved IPv6 inter-communication and opened IPv6 international outlets. By 2025, China is expected to be the country with the most IPv6 users in the world.

1.2 Increasingly Diverse Patterns of Internet Technological Innovation

Many countries put sustainable investment into information technologies. Technological innovation in the world has achieved exponential growth, which activates a new leap of economic and social development. In 2019, the USA remained the leading role in global information technology and industry. As Price Waterhouse Coopers (PwC) analyzes, among global top ten innovative corporations, eight are American corporations related to Internet industry and information technology (Apple, Amazon, Alphabet, Microsoft, Tesla, Facebook, Intel, and Netflix).[2] Meanwhile, there are increasingly diverse patterns of global technological innovation landscape. Most countries have constantly increased the proportion of business in Internet and the investment into venture capital, and formulated relevant strategic plans successively. They compete and rival in critical fields of new technologies such as intelligence (AI) and quantum computing. For instance, in 2018, the USA published Rise of the Machines: Artificial Intelligence and Its Growing Impact of on US Policy, EU published Coordinated Plan on Artificial Intelligence, France formulated Artificial Intelligence Strategy, and Germany published Key Points for a Strategy on Artificial Intelligence.

[1] Qualcomm: 5G Economy. 2017 5G Summit, February 22, 2017.

[2] PwC Strategy: The 2018 Global Innovation 1000 Study, November 1, 2018.

1.3 Thriving Global Digital Economy with Highlights

Presently, world economy is facing a crucial turn of growth drivers, and digital economy proves a strong driver for the sustainable economic and social transformation. The proportion of digital economy in national GDP shows upward and fuels GDP growth remarkably. The United Nations Conference on Trade and Development estimated that the size of digital economy accounted for 4.5%–15.5% of world GDP. Digital economy and real economy further integrate and develop, integration across industries and vertical integration grow vigorously, and digital industrialization and industrial digitalization proceed steadily. Global electronics and information manufacturing industry is in constant growth. Service industry and digital technology deeply integrate, giving birth to new business modes and forms such as sharing economy and platform economy. Global industrial Internet platform industry is developing rapidly. Digital transformation of manufacturing industry that is driven by new-generation information technologies such as big data, cloud computing, and artificial intelligence presents great potential. Industrial Internet, fintech, AI applications, and intelligent city become highlights in development. Digital economy plays a pivotal role in improving international competitiveness. It is a common choice for all countries to accelerate the transformation of digital technology dividends. Digital technology helps more developing economies and small and medium-sized enterprises to participate in global value chain; however, it adversely affects employment in developing countries. This deserves attention.

1.4 New Technologies' Profound Influence on the Development of Internet Media

In 2019, Internet media industry developed rapidly, with further increase of users. Statistics shows that the number of social media users worldwide increased to 3.5 billion in early 2019, with global penetration rate up to 46%.[3] Among them, mobile Internet media users were the mainstream. Social media became an important platform to obtain news. The number of users who get access to news and information via Internet and social media increased steadily. Internet media companies and traditional news industry are seeking new modes of win-win cooperation. Digital news subscription may become major income source for news industry in the future. Global video on demand (VOD) market expanded, leading platforms displayed prominent advantage, digital music industry developed steadily, and streaming music made

[3]Data based on the number of active users per month on the most active social platforms in each country. Data source: Hootsuite, July 28, 2019.

great profit. The application of new technologies deeply affected the production and organization of Internet media. 5G technology effectively raised the efficiency of data production and communication and deeply improved user's experience. Cloud technology changed media's production system and promoted resource sharing and utilization. Artificial intelligence further optimized media's value chain and enhanced the efficiency and communication of content production. At the same time, social media, especially personal instant messaging tools, became headstream of online rumors and false information. The monopoly of Internet media platform brought challenges to governmental management, industrial development, and technological progress.

1.5 Increasingly Prominent Global Cybersecurity Threats

Cybersecurity incidents, such as data leakage, cyberattacks, and ransomware, occur frequently. Emerging technologies such as artificial intelligence, Internet of things, and cloud computing advance constantly, yet new cybersecurity threats iterate and evolve quickly. Security threats such as cyberattacks, misuse of technology, and intelligent destructive weapons aiming at critical information infrastructure evoke awareness of all countries. In August 2018, Venezuela's president was attacked by unmanned aerial vehicle bomb in public. It is the first global terrorist activity using artificial intelligence products. Facing the increasingly severe situation of cybersecurity, many countries have equated cybersecurity with national security at the same strategic level and took various protective measures, including perfecting systems, establishing special institutions, strengthening personnel training, and improving cybersecurity monitoring mechanism, in order to enhance the comprehensive protection capacity of cybersecurity. Competition in cyberspace across countries poses challenges to global cybersecurity. With the militarization of global cyberspace intensified, strategic stability in cyberspace urgently needs to be standardized, and international cooperation needs to be strengthened and deepened.

1.6 Historic Transition Period of Global Cyberspace Governance

Now, cyberspace and national governance deeply integrate, and geopolitical conflicts spread to cyberspace gradually. Cyberspace governance is in a historic transition period when multilateral/multi-party governance parallels. Especially, as the competition between leading powers intensifies, the vulnerability and uncertainty of cyberspace international rules become more prominent. Confronted with complicated international environment, state actors and non-state actors repeatedly express their increasing appeal to participate in the governance and to actively set up new rules for cyberspace governance in order to build a good global operating environment. Over the past year, international governance platforms, represented by the United Nations, continuously promoted international cyberspace governance and made progress in the fields of cyberspace rule-making, digital economy development and cybersecurity. Accountability system reform of Internet Corporation for Assigned Names and Numbers (ICANN) advanced by trial. Negotiation of global community in the post-Internet Assigned Numbers Authority (IANA) time continued to be enhanced. The Global Commission on the Stability of Cyberspace (GCSC) and some other platforms strengthened research on rule-making and proposed rules and suggestions. Traditional international organizations, such as Group of 20 (G20), BRICS, and Shanghai Cooperation Organization (SCO), accelerated their participation in cyberspace governance. In June 2019, G20 Summit was held in Osaka, Japan, when 24 countries, including China, the USA, Japan, and the UK, reached a consensus on international data flow and principles in artificial intelligence, to deal with the challenges related to the protection of privacy and data, intellectual property rights (IPR), and security.

2 Evaluation and Analysis of Internet Development in Global Representative Countries in 2019

In 2017, World Internet Development Report firstly established Global Internet Development Index (GIDI) system. In 2019, GIDI System selected and analyzed 48 representative countries on the five continents to present the latest world Internet development. The 48 representative countries are listed as below:

America: The USA, Canada, Brazil, Argentina, Mexico, Chile, Cuba.
Asia: China, Japan, South Korea, Indonesia, India, Saudi Arabia, Turkey, the United Arab Emirates, Malaysia, Singapore, Thailand, Israel, Kazakhstan, Vietnam, Pakistan, Iran.
Europe: The UK, France, Germany, Italy, Russia, Estonia, Finland, Norway, Spain, Switzerland, Denmark, Netherlands, Portugal, Sweden, Ukraine, Poland, Ireland, Belgium.
Oceania: Australia, New Zealand.
Africa: South Africa, Egypt, Kenya, Nigeria, Ethiopia.

2.1 Index Construction

Global Internet Development Index (GIDI) comprehensively measures and reflects a country's Internet development according to six aspects: infrastructure, innovation capacity, industrial development, Internet application, cybersecurity, and cyberspace governance. In the past two years, indicator system consisted of six first-level indicators, 12–15 second-level indicators and many third-level indicators, as shown in Fig. 1. Based on the researches in the previous two years and considering the availability of metadata of various indicators, Global Internet Development Index (GIDI) in 2019 maintained six first-level indicators and revised second-level indicators (14) and third-level indicators (31).

As smart phones are widely used and mobile applications are developing rapidly, creation of mobile applications was added as a third-level indicator. To precisely present the level of consumer's shopping in Internet in different countries, the percentage of online shopping in domestic retail was adopted to replace the income of online shopping market. To reflect company's capacity in using Internet, the use of Internet in business-to-customer (B2C) transaction was added as a third-level indicator. To take providing help to other countries in Internet construction as an evaluation criterion, leading or participating in cyber-capacity building was added as a third-level indicator. Additionally, cybersecurity commitment indicator was deleted considering the availability of data.

2.2 Assignment of Weights

Infrastructure, innovation capacity, industrial development, Internet application, cybersecurity, and cyberspace governance are main factors that affect Internet development. The weights of these factors are basically consistent with those in 2018, and data sources used are modified. See Table 1 for Global Internet Development Index (GIDI) system and indicator description.

2.3 Analysis of Results

Through the calculation of indicators, the scores of Internet Development Index in 48 countries were obtained (see Table 2). As it signals, China and the USA take the lead, yet the gap between China and the USA in overall technology strength is still very big. European developed countries, such as the UK, France, and Germany, maintain a high level of Internet development. Latin America and Sub-Saharan Africa are still increasing efforts on development.

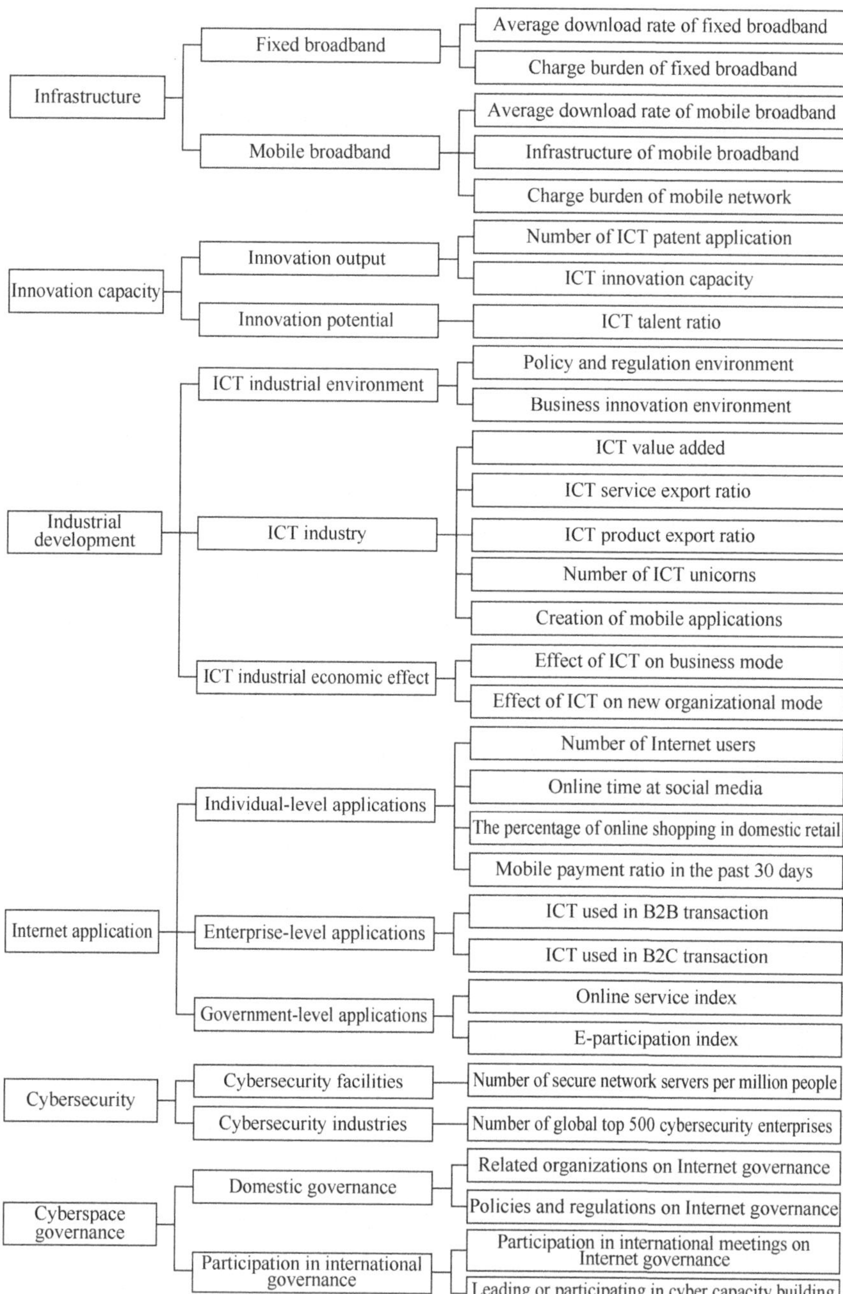

Fig. 1 Global internet development index (GIDI) system

Table 1 Global internet development index (GIDI) system and indicator description

First-level indicators	Second-level indicators	Third-level indicators	Indicator description	Data sources
1. Infrastructure 10%	1.1 Fixed broadband	1.1.1 Average download rate of fixed broadband	Reflecting the average download rate of fixed broadband users over a given period of time in different countries	Statistical data of Global Digital Reports (by Global Web Index, etc.), 2018
		1.1.2 Charge burden of fixed broadband	Reflecting the ratio of fixed broadband charge in GNI	Database of International Telecommunication Union, 2017
	1.2 Mobile broadband	1.2.1 Average download rate of mobile broadband	Reflecting the average download rate of mobile broadband users over a given period of time in different countries	Statistical data of Global Digital Reports (by Global Web Index, etc.), 2018
		1.2.2 Infrastructure of mobile network	Reflecting the infrastructure construction of mobile network in different countries	Statistical data of Global Digital Reports (by Global Web Index, etc.), 2018
		1.2.3 Charge burden of mobile network	Reflecting the ratio of mobile network charge in GNI	Database of International Telecommunication Union, 2017
2. Innovation capacity 20%	2.1 Innovation output	2.1.1 Number of ICT patent application	Reflecting the level and capacity of ICT patent application in different countries	Database of Organization for Economic Cooperation and Development, 2016
		2.1.2 ICT innovation capacity[a]	Reflecting the innovation level and capacity of ICT industry in different countries	Statistical data of World Economic Forum, 2018
	2.2 Innovation potential	2.2.1 ICT talent ratio	Reflecting the ratio of ICT talent in the population of different countries	Database of International Labor Organization, 2018

(continued)

Table 1 (continued)

First-level indicators	Second-level indicators	Third-level indicators	Indicator description	Data sources
3. Industrial development 20%	3.1 ICT industrial environment	3.1.1 Policy and regulation environment[a]	Reflecting the policy and regulation environment in ICT industry in different countries	Statistical data of World Economic Forum, 2018
		3.1.2 Business innovation environment[a]	Reflecting the business environment in ICT industry in different countries	Statistical data of World Economic Forum, 2018
	3.2 ICT industry	3.2.1 ICT value added	Reflecting the ICT value added in different countries	Database of the United Nations, 2017
		3.2.2 ICT service export ratio	Reflecting the ratio of ICT service export scale in that of domestic service in different countries	Statistical data from World Development Indicators of the World Bank, 2017
		3.2.3 ICT product export ratio	Reflecting the ratio of ICT product export scale in that of domestic products in different countries	Statistical data from World Development Indicators of the World Bank, 2017
		3.2.4 Number of ICT unicorns	Reflecting the number of ICT corporations with the market value over $1 billion in different countries	Statistical data of CB Insights, 2018
		3.2.5 Creation of mobile applications	Reflecting the creation of mobile applications in different countries	Statistical data of World Intellectual Property Organization, 2018

(continued)

Table 1 (continued)

	3.3 ICT industrial economic effect	3.3.1 Effect of ICT on business mode	Reflecting the improvement of business mode with ICT in different countries	Statistical data of World Economic Forum, 2017
		3.3.2 Effect of ICT on new organizational mode	Reflecting the improvement of organizational mode with ICT in different countries, like virtual team and telework	Statistical data of World Economic Forum, 2017
4. Internet application 30%	4.1 Individual-level applications	4.1.1 Number of Internet users	Reflecting the total number of Internet users in different countries	Statistical data of Global Digital Reports (by Global Web Index, etc.), 2018
		4.1.2 Online time at social media	Reflecting the online time at social media in different countries	Statistical data of Global Digital Reports (by Global Web Index, etc.), 2018
		4.1.3 The percentage of online shopping in domestic retail	Reflecting the percentage of consumer's online shopping in domestic retail in different countries	Statistical data of Global Digital Reports (by Global Web Index, etc.), 2018
		4.1.4 Mobile payment ratio in the past 30 days	Reflecting the ratio of mobile payment in different countries	Statistical data of Global Digital Reports (by Global Web Index, etc.), 2018
	4.2 Enterprise-level applications	4.2.1 ICT used in B2B transaction	Reflecting enterprise's capacity in using ICT in B2B transaction in different countries	Statistical data of World Economic Forum, 2017
		4.2.2 ICT used in B2C transaction	Reflecting the role of Internet in e-commerce	Statistical data of World Economic Forum, 2017

(continued)

Table 1 (continued)

	4.3 Government-level applications	4.3.1 Online service index	Reflecting the quality of online service provided by government websites in different countries	Statistical data of the United Nations, 2017
		4.3.2 E-participation index	Reflecting the online communication between citizens and government in different countries	Statistical data of the United Nations, 2017
5. Cybersecurity 10%	5.1 Cybersecurity facilities	5.1.1 Number of secure network servers per million people	Reflecting the number of secure network servers per 1 million people in different countries	Statistical data from database of the World Bank, 2018
	5.2 Cybersecurity industries	5.2.1 Number of global top 500 cybersecurity enterprises	Reflecting the number of Cybersecurity 500 in different countries	List of Cybersecurity 500 published by Cybersecurity Ventures, 2018
6. Cyberspace governance 10%	6.1 Internet governance	6.1.1 Related organizations on Internet governance	Reflecting related organizations in Internet governance in different countries, in such specific affairs as policy, security, protection of critical information infrastructure, CERT, criminal, and protection of consumer	Referring to achievement of foreign researches and inviting experts and scholars in relevant fields to conduct comprehensive evaluation
		6.1.2 Policies and regulations on Internet governance	Reflecting Internet affairs or the making of policies and regulations related to ISP in different countries	Referring to achievement of foreign researches and inviting experts and scholars in relevant fields to conduct comprehensive evaluation

(continued)

Table 1 (continued)

	6.2 Participation in international governance	6.2.1 Participation in international meetings on Internet governance	Reflecting participation in international conferences on cyberspace in different countries, including bilateral meetings, multilateral meetings and other forums	Referring to achievement of foreign researches and inviting experts and scholars in relevant fields to conduct comprehensive evaluation
		6.2.2 Leading or participating on cyber-capacity construction	Reflecting helping others in cyber-capacity construction, including technological assistance, policy guidance, or project training	Referring to achievement of foreign researches and inviting experts and scholars in relevant fields to conduct comprehensive evaluation

Note [a]indicates the data calculated based on the database of World Economic Forum, 2011–2016.

Table 2 Scores of internet development index in 48 countries

No.	Countries	Score	Rank
1	The USA	63.86	1
2	China	53.03	2
3	South Korea	49.63	3
4	The UK	49.02	4
5	France	48.49	5
6	Finland	48.28	6
7	Sweden	47.83	7
8	Singapore	47.71	8
9	Germany	47.50	9
10	Japan	47.21	10
11	Norway	46.47	11
12	Canada	46.28	12
13	Switzerland	46.17	13
14	Israel	45.96	14
15	Denmark	45.66	15
16	Netherlands	44.59	16
17	Russia	44.48	17
18	Australia	44.43	18
19	Estonia	44.08	19
20	Spain	43.68	20
21	Ireland	42.94	21
22	New Zealand	42.85	22
23	India	42.81	23
24	Poland	42.67	24
25	Italy	42.02	25
26	The United Arab Emirates	41.07	26
27	Brazil	40.60	27
28	Belgium	40.60	27
29	Turkey	40.54	28
30	Malaysia	40.42	29
31	Vietnam	40.16	30
32	Thailand	39.22	31
33	Indonesia	38.56	32
34	South Africa	38.43	33
35	Portugal	37.76	34
36	Mexico	36.96	35
37	Ukraine	36.93	36

(continued)

Table 2 (continued)

38	Argentina	36.49	37
39	Saudi Arabia	36.29	38
40	Egypt	34.92	39
41	Chile	32.70	40
42	Iran	32.54	41
43	Pakistan	30.61	42
44	Kenya	30.35	43
45	Kazakhstan	29.83	44
46	Nigeria	29.54	45
47	Ethiopia	25.68	46
48	Cuba	23.39	47

2.3.1 Information Infrastructure Was Further Optimized and Upgraded, with Big Difference in Cyber-Capacity Construction Among 48 Countries

In 48 countries, governments took an active attitude in constructing information infrastructure, which continuously improved network quality and internet speed. The evaluation results indicate that, in countries with small land area and more developed economy, such as Singapore, Norway, Sweden, Switzerland, Denmark, and South Korea, there is advantage in information infrastructure construction. Yet, in countries with vast land, like the USA, China and India, there is unbalanced development of infrastructure construction in different regions, which affects the average level of information infrastructure in these countries.

Information infrastructure such as 5G and IPv6 grew rapidly, and mobile broadband is evolving toward 5G age. In The Mobile Economy 2019, it is predicted that as web publishing and compatible devices increased in 2019, 5G will be connected to 1.4 billion users by 2025. IPv6 commercial deployment was extensively carried out, with stable growth in penetration rate and traffic. According to the statistical data of Asia Pacific Network Information Center (APNIC), global deployment rate of IPv6 had reached 22.84% in late June, 2019. In India, the USA, and Belgium, it exceeded 50%.

Currently, there is big difference in cyber-construction capacity among 48 countries. Countries and regions in North America, Europe, and Asia are endeavoring to develop 5G. According to a test of OpenSignal, among the eight major countries and regions that have started the commercial uses of 5G, the USA displays higher level and provides better user's experience in 5G construction, with the maximum downlink rate in the real world reaching 1815 Mb/s, 2.7 times of 4G peak. Switzerland, South Korea, Australia, the United Arab Emirates, Italy, Spain, and the UK closely follow it. However, Sub-Saharan African countries are still using 2G or 3G, and some have assigned 4G spectrums to telecom operators. See Table 3 for scores of Information infrastructure in 48 countries.

Table 3 Scores of information infrastructure in 48 countries

No.	Countries	Score	Rank
1	Singapore	6.00	1
2	Norway	5.17	2
3	Sweden	5.09	3
4	Switzerland	4.73	4
5	Denmark	4.68	5
6	France	4.65	6
7	Canada	4.50	7
8	South Korea	4.44	8
9	Russia	4.31	9
10	Australia	4.31	9
11	Belgium	4.22	10
12	The USA	4.10	11
13	Netherlands	4.09	12
14	Poland	4.07	13
15	Israel	3.91	14
16	Germany	3.90	15
17	Finland	3.89	16
18	Spain	3.86	17
19	The UK	3.85	18
20	The United Arab Emirates	3.83	19
21	Japan	3.70	20
22	New Zealand	3.60	21
23	Ireland	3.57	22
24	Estonia	3.55	23
25	Italy	3.38	24
26	Portugal	3.30	25
27	Turkey	3.25	26
28	China	3.23	27
29	Malaysia	3.03	28
30	Iran	3.02	29
31	Egypt	2.89	30
32	Kazakhstan	2.83	31
33	Ukraine	2.74	32
34	Saudi Arabia	2.69	33
35	Chile	2.61	34

(continued)

Table 3 (continued)

36	Thailand	2.50	35
37	Argentina	2.44	36
38	Vietnam	2.43	37
39	Mexico	2.42	38
40	South Africa	2.34	39
41	Brazil	2.27	40
42	India	1.86	41
43	Indonesia	1.85	42
44	Nigeria	1.65	43
45	Kenya	1.60	44
46	Pakistan	1.57	45
47	Ethiopia	1.41	46
48	Cuba	0.96	47

2.3.2 Internet Technological Innovation Capacity Is Growing Steadily in 48 Countries, Constructing Innovative Countries

Innovation capacity is an important indicator that presents the potential in Internet future development of a country. The USA takes the lead in global information technology and industry. It owns R&D-intensive and innovative high-tech corporations and universities, ranking the first in innovation quality in the world. At the same time, the pattern of global scientific and technological innovation is presenting a trend of multi-polarization.

European markets look small and fragmented, but they keep ahead in scientific and technological innovation. Dublin of Ireland, Grenoble of France, Berlin of Germany, Krakow of Poland, and Tallinn of Estonia are all hailed as "Silicon Valley of Europe." With a great deal of universities and institutions in this area, they enjoy rich R&D resources. In ICT patent application and ICT talent quantity, the UK, France, and Germany make striking achievement. In particularly, Switzerland, Sweden, Netherlands, Finland, and Denmark ranked top ten in Global Innovation Index 2019, entitled as the most innovative countries.

With the rapid development of China, India, and South Korea, Asian countries are constantly improving their scientific and technological innovation capacity, and especially India is prominent. According to the statistics of World Intellectual Property Organization (WIPO) in March 2019, China, Japan, and South Korea ranked the second, third, and fifth among top ten countries in PCT patent application quantity. India grew fastest, at a rate of 27.2% and ranking among the top 15. See Table 4 for scores of Internet innovation capacity in 48 countries.

Table 4 Scores of internet innovation capacity in 48 countries

No.	Countries	Score	Rank
1	The USA	9.10	1
2	China	8.96	2
3	Japan	8.90	3
4	South Korea	8.40	4
5	Germany	8.22	5
6	Sweden	7.97	6
7	The UK	7.92	7
8	France	7.83	8
9	Canada	7.71	9
10	Israel	7.67	10
11	Netherlands	7.49	11
12	Finland	7.48	12
13	India	7.40	13
14	Switzerland	7.39	14
15	Australia	7.25	15
16	Italy	7.12	16
17	Singapore	7.06	17
18	Belgium	6.99	18
19	Ireland	6.99	18
20	Russia	6.98	19
21	Denmark	6.89	20
22	Spain	6.88	21
23	Malaysia	6.78	22
24	Turkey	6.67	23
25	Norway	6.65	24
26	Brazil	6.51	25
27	Poland	6.44	26
28	South Africa	6.29	27
29	Saudi Arabia	6.23	28
30	Mexico	6.21	29
31	New Zealand	6.21	29
32	Portugal	6.20	30
33	Ukraine	6.13	31
34	The United Arab Emirates	5.97	32
35	Estonia	5.62	33

(continued)

Table 4 (continued)

36	Chile	5.50	34
37	Thailand	5.45	35
38	Iran	5.39	36
39	Egypt	5.21	37
40	Kenya	5.05	38
41	Argentina	5.02	39
42	Indonesia	5.00	40
43	Nigeria	4.99	41
44	Pakistan	4.99	41
45	Kazakhstan	4.53	42
46	Vietnam	4.51	43
47	Ethiopia	4.44	44
48	Cuba	3.89	45

2.3.3 48 Countries are Actively Developing Internet Industry and Unicorns Are Mainly in China and the USA

Internet industry keeps growing rapidly, forming notable coordinated development with real economy. New economic growth drivers play more important role. China and the USA became the ideal land for Internet newly founded enterprises with global advantage. In Global Unicorns List 2019 issued by CB Insights, 326 companies from China and the USA were included, accounting for 76.3% in the world. The USA boasts most unicorns (159), accounting for 48%. China ranks second in the number of unicorns (92), accounting for 28%. 17 unicorns (5%) come from the UK and 13 unicorns (4%) come from India, ranking the third and fourth, respectively. The number of Indian unicorns grew fast and its ranking rose gradually. In spite that European countries take a backward step in Internet newly founded enterprises and new business modes, Germany, Switzerland, Finland, and Estonia remain ahead in ICT service export. Asia-Pacific countries, such as China, South Korea, Malaysia, and Singapore, have advantage in ICT product export.

With the increasing use of smart phones in the world and the commercial uses of 5G, mobile apps are getting diversified and mobile app economy is expected to thrive again. App Annie, a market research firm, estimates that global economic scale of mobile apps will reach $101 billion in 2020. In larger markets, the USA, China, and Japan, mobile apps income increases the most. In fast-growing markets, India, Indonesia, Mexico, and Argentina, there will be plenty of opportunities for mobile app economy. See Table 5 for scores of Internet industrial development in 48 countries.

Table 5 Scores of internet industrial development in 48 countries

No.	Countries	Score	Rank
1	The USA	18.00	1
2	China	15.68	2
3	Israel	14.43	3
4	Finland	14.27	4
5	South Korea	12.88	5
6	Sweden	12.80	6
7	Estonia	12.76	7
8	Denmark	12.36	8
9	Switzerland	11.97	9
10	Singapore	11.94	10
11	Ireland	11.81	11
12	The UK	11.68	12
13	Vietnam	11.59	13
14	France	11.56	14
15	Ukraine	11.22	15
16	Netherlands	11.04	16
17	Canada	10.92	17
18	New Zealand	10.88	18
19	Russia	10.84	19
20	Japan	10.79	20
21	India	10.68	21
22	Poland	10.65	22
23	Norway	10.62	23
24	Australia	10.56	24
25	Germany	10.47	25
26	Spain	10.27	26
27	Brazil	10.06	27
28	Turkey	9.94	28
29	The United Arab Emirates	9.30	29
30	South Africa	9.29	30
31	Malaysia	9.23	31
32	Pakistan	9.16	32
33	Italy	8.87	33
34	Argentina	8.85	34
35	Indonesia	8.72	35
36	Thailand	8.69	36
37	Portugal	8.31	37

(continued)

Table 5 (continued)

38	Belgium	8.20	38
39	Mexico	8.12	39
40	Saudi Arabia	7.78	40
41	Chile	7.69	41
42	Egypt	7.66	42
43	Nigeria	7.59	43
44	Kenya	7.50	44
45	Kazakhstan	7.49	45
46	Iran	6.06	46
47	Ethiopia	5.85	47
48	Cuba	5.68	48

2.3.4 Individual-level Internet Applications in Countries with Large number Internet Users Are Advantageous, and Enterprise-level and Government-level Applications in Developed Countries Are Better

Individual, enterprise, and government are three main bodies of Internet applications. In individual-level applications, countries with a large number of Internet users have obvious advantages. For example, in the USA, China, India, and Indonesia, social media, online shopping, and ride-hailing are widely used services. According to the statistical data of Global Digital Reports 2019, global Internet users spend 6 h and 42 min online per day on the average. Most time is on social media and reaches 2 h and 16 min averagely, especially for Internet users in South American countries such as Brazil, Argentina, Mexico, and Chile.

In enterprise-level Internet applications, developed countries or regions such as the USA and Europe are developing at a high level, yet many developing countries in Asia, South America, and Africa fall behind. In terms of enterprise's strength, the USA and European countries (e.g., Germany) remain powerful. Oracle, Salesforce, Workday, and Servicenow are leading users of enterprise-level applications. Microsoft and Amazon also have a large percentage of enterprise-level businesses. In the field of industrial robotics, ABB from Switzerland, Kuka from Germany, and FANUC and Yaskawa from Japan almost monopolize global relevant markets. Depending on the supports of these enterprises and strong industrial foundation, the US and European enterprises display top level in enterprise-level Internet applications, far ahead of developing countries.

In government-level applications, in general, there is a higher-level e-government in more advanced countries. European countries play a leading role in the development of global e-government, with more and more countries providing mobile applications and other online services. In some American and Asian developed countries, such as the USA, South Korea, Japan, and Singapore, e-governments rank top in the world. In Africa, however, e-government commonly remains at a low level. See Table 6 for scores of Internet applications in 48 countries.

Table 6 Scores of internet applications in 48 countries

No.	Countries	Score	Rank
1	China	13.90	1
2	The UK	13.87	2
3	The USA	13.86	3
4	Norway	13.72	4
5	Thailand	13.54	5
6	Indonesia	13.52	6
7	Germany	13.52	6
8	Finland	13.42	7
9	France	13.06	8
10	South Korea	13.05	9
11	India	12.86	10
12	Netherlands	12.85	11
13	Vietnam	12.64	12
14	Japan	12.53	13
15	Sweden	12.48	14
16	Poland	12.46	15
17	Malaysia	12.44	16
18	Spain	12.34	17
19	Italy	12.34	17
20	Denmark	12.32	18
21	Brazil	12.20	19
22	Canada	12.04	20
23	Singapore	12.03	21
24	Turkey	12.02	22
25	Australia	11.98	23
26	Mexico	11.86	24
27	Belgium	11.85	25
28	New Zealand	11.85	25
29	The United Arab Emirates	11.73	26
30	Switzerland	11.65	27
31	Ireland	11.46	28
32	Russia	11.31	29
33	Argentina	11.27	30
34	Estonia	11.05	31
35	South Africa	10.69	32
36	Portugal	10.59	33
37	Egypt	10.02	34

(continued)

Table 6 (continued)

38	Saudi Arabia	10.02	35
39	Israel	9.84	36
40	Iran	9.43	37
41	Chile	8.70	38
42	Ukraine	8.10	39
43	Kenya	8.00	40
44	Nigeria	6.67	41
45	Kazakhstan	6.28	42
46	Pakistan	6.27	43
47	Ethiopia	5.55	44
48	Cuba	4.34	45

2.3.5 The USA Takes Leading Role in Global Cybersecurity, and EU Highlights Protection of Privacy

The USA attaches much importance to constructing a cybersecurity system that can defend and attack. Besides, American cybersecurity corporations possess powerful strength. According to Global Cybersecurity Innovation Top 500 released by Cybersecurity Ventures, a research platform, in May 2018, 358 American corporations were listed, ranking the first and remaining dominant. With 42 corporations listed, Israel ranked the second and revealed strong capacity in cybersecurity industry capacity.

EU underlines the protection of privacy and data and issued General Data Protection Regulation (GDPR). It becomes an important reference for other countries when they formulate privacy policies. In this way, EU succeeded in standardizing rules on data protection in its member states, which provided more stable legal anticipation for individuals and enterprises. However, some studies revealed that over the past year, in practice, GDPR was proved to be too complex for consumers to understand and for enterprises to observe. This impeded user's online access and resulted in regulators' resource shortage. Even worse, this damnified European technological newly founded enterprises and harmed technological development. See Table 7 for scores of cybersecurity capacity in 48 countries.

2.3.6 The Concept of Government-led Internet Governance Received More Supporters, and the Mode of Multilateral Participation Was Gradually Accepted

In the time of digital economy, more extensive national policies and regulatory framework are necessary and to balance regulation and innovation become a hard task. Admittedly, on making international rules in cyberspace, wide divergences exist among countries, and the dispute between multilateralism and multi-interest-related party mode continues. Nevertheless, international community further enhanced their consensus on cyberspace governance, and more countries recognized the importance of cyber-sovereignty, with governments playing more prominent role in Internet governance. A report of Washington Think Tank Center for Global Development

Table 7 Scores of cybersecurity capacity in 48 countries

No.	Countries	Score	Rank
1	The USA	9.30	1
2	Israel	3.46	2
3	The UK	3.15	3
4	Germany	2.86	4
5	France	2.85	5
6	Sweden	2.84	6
7	Switzerland	2.84	6
8	Singapore	2.80	7
9	Denmark	2.78	8
10	Netherlands	2.77	9
11	Finland	2.77	9
12	Ireland	2.75	10
13	Estonia	2.74	11
14	Japan	2.74	11
15	Canada	2.74	11
16	Australia	2.73	12
17	Spain	2.73	12
18	New Zealand	2.72	13
19	Norway	2.72	13
20	Italy	2.71	14
21	Portugal	2.71	14
22	China	2.71	14
23	South Africa	2.70	15
24	Belgium	2.70	15
25	Chile	2.69	16
26	Poland	2.69	16
27	Russia	2.68	17
28	South Korea	2.68	17
29	Malaysia	2.68	17
30	Ukraine	2.67	18
31	Turkey	2.66	19
32	Argentina	2.64	20
33	Brazil	2.64	21
34	Vietnam	2.63	22
35	The United Arab Emirates	2.63	22
36	Indonesia	2.63	22
37	Kazakhstan	2.63	22

(continued)

Table 7 (continued)

38	India	2.62	23
39	Thailand	2.60	24
40	Iran	2.57	25
41	Nigeria	2.57	25
42	Mexico	2.56	26
43	Saudi Arabia	2.56	26
44	Pakistan	2.54	27
45	Kenya	2.50	28
46	Egypt	2.50	28
47	Cuba	2.45	29
48	Ethiopia	2.37	30

indicates that more and more countries, including some Western African countries and Vietnam, are taking China as an example and learning China's Internet regulatory mode. See Table 8 for scores of Internet governance in 48 countries.

Table 8 Scores of internet governance in 48 countries

No.	Countries	Score	Rank
1	The USA	9.50	1
2	China	8.55	2
3	Japan	8.55	2
4	The UK	8.55	2
5	France	8.55	2
6	Germany	8.55	2
7	Canada	8.36	3
8	Russia	8.36	3
9	Estonia	8.36	3
10	South Korea	8.17	4
11	Singapore	7.88	5
12	The United Arab Emirates	7.60	6
13	Italy	7.60	6
14	Norway	7.60	6
15	Spain	7.60	6
16	Switzerland	7.60	6
17	Australia	7.60	6
18	New Zealand	7.60	6
19	India	7.40	7

(continued)

Table 8 (continued)

20	South Africa	7.12	8
21	Saudi Arabia	7.02	9
22	Brazil	6.93	10
23	Indonesia	6.83	11
24	Egypt	6.64	12
25	Israel	6.64	12
26	Denmark	6.64	12
27	Portugal	6.64	12
28	Sweden	6.64	12
29	Belgium	6.64	12
30	Thailand	6.45	13
31	Finland	6.45	13
32	Vietnam	6.36	14
33	Netherlands	6.36	14
34	Poland	6.36	14
35	Ireland	6.36	14
36	Argentina	6.26	15
37	Malaysia	6.26	15
38	Cuba	6.07	16
39	Kazakhstan	6.07	16
40	Pakistan	6.07	16
41	Ukraine	6.07	16
42	Nigeria	6.07	16
43	Ethiopia	6.07	16
44	Iran	6.07	16
45	Turkey	5.98	17
46	Mexico	5.79	18
47	Kenya	5.69	19
48	Chile	5.50	20

3 Internet Development in Some Typical Countries

As the scores of Internet Development Index in 48 countries reveal, developed countries in North America, Europe, and Asia maintain a high level of Internet development. Latin America and Sub-Saharan African developing countries are accelerating development. Among them, Internet development in the USA, China, South Korea, France, Germany, Israel, Russia, India, Brazil, and South Africa are typical. The analyses on the selected ten countries are as below.

3.1 The USA

As a world Internet power, the USA keeps the leading role in global Internet technological innovation and development. In the list of World Internet Development Index, the USA ranks the first, and remains top in innovation capacity, industrial development, cybersecurity, and cyberspace governance, with strong strength in all fields. In Internet application, the USA is second only to China and the UK and ranks the third. In infrastructure, with digital divide unsolved, the USA ranks 12th. See Fig. 2 for Internet Development Index of the USA.

The US government concentrates on information infrastructure construction. Telecom and Internet operators Verizon and AT&T continuously increased investment in fiber optic networks. Predictably, AT&T will provide 12.5 million homes with fiber to the premises (FTTP) service with 1Gb/s by 2019. The USA consolidated 5G deployment and promoted "5G FAST plan." American spectrum regulators have completed spectrum auctions suitable for 5G deployment. Twenty-one states have enacted laws and regulations to accelerate the deployment of small base stations.

The USA possesses obvious advantages in capital investment, number of patents and technological publications, and talent construction. Particularly, in the development of industrial clusters, the USA owns top innovation clusters such as Silicon Valley.

American Internet corporations became the bellwether of global Internet development. In Top 10 Tech Companies 2019 released by the Forbes, eight comes from

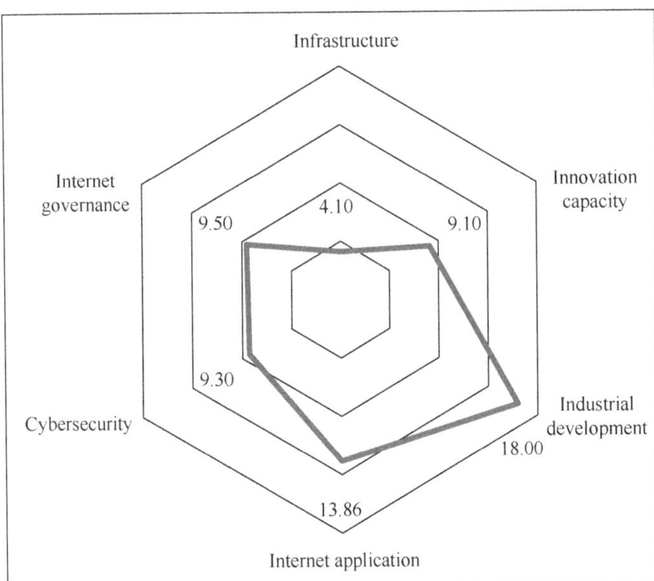

Fig. 2 Internet development index of the USA

the USA, including Apple, Microsoft, Alphabet, Intel, IBM, Facebook, Cisco, and Oracle.

The USA government attaches great importance to cybersecurity and formulated and issued a great deal of policies. In May 2018, Department of Homeland Security (DHS) published Cybersecurity Strategy, which provided a framework for the USA to implement cybersecurity responsibilities over the next five years. The US Department of Energy published Multiyear Plan for Energy Sector Cybersecurity to reduce the risk of cyber-incidents to American energy. The current National Security Strategy of the USA also places great emphasis on cybersecurity. In August 2019, National Institute of Standards and Technology (NIST) published a draft of IoT Device Cybersecurity Capability Core Baseline, which aimed to reduce the security risk of Internet of things (IoT).

3.2 China

In Internet development, China is second only to the USA. Notably, in Internet application, China ranks the first; in innovation capacity, China ranks the second; and in industrial development, China ranks the second. China's plan in IT patent contributes to the rapid rise of innovation capacity, with the advantage in patent quantity expanded. In Internet governance, China ranks the second. Internet governance system is improved gradually and accumulated experiences in practice. Yet, in infrastructure (ranking the 28th) and cybersecurity (ranking the 20th together), China must make efforts to improve. See Fig. 3 for Internet Development Index of China.

Chinese government has constantly strengthened information infrastructure construction, accelerated the practice of "Broadband China" strategy, and actualized "Boosting Internet Speed and Lowering Internet Charges" policy. In fiber broadband, China takes the lead in the world, with more than 90% broadband users using optical fiber access. By the end of December 2018, Internet broadband access ports had reached 886 million, with a net increase by 110 million than that by the end of 2017. In mobile Internet, commercial uses of 5G speeds up. In particular, 5G commercial license has been officially issued, marking a great leap in mobile Internet. Internet of things (IoT) is expected to make explosive growth and benefit smart home and smart car first. And a tremendous smart device market will be activated.

In innovation capacity, China has deeply implemented innovation-driven development strategy and accelerated the development of cutting-edge technologies such as artificial intelligence, quantum computing, and neural network chips. In many fields, such as network technology, mobile chips, intelligent terminals, cloud computing and big data, China has made breakthroughs. According to a statistical report of World Intellectual Property Organization (WIPO), China ranks top in three IPR applications in the world, patent, trademark, and industrial design. China plays a leading role in 5G international standard. In light of the statistics published by IPLytics, a German

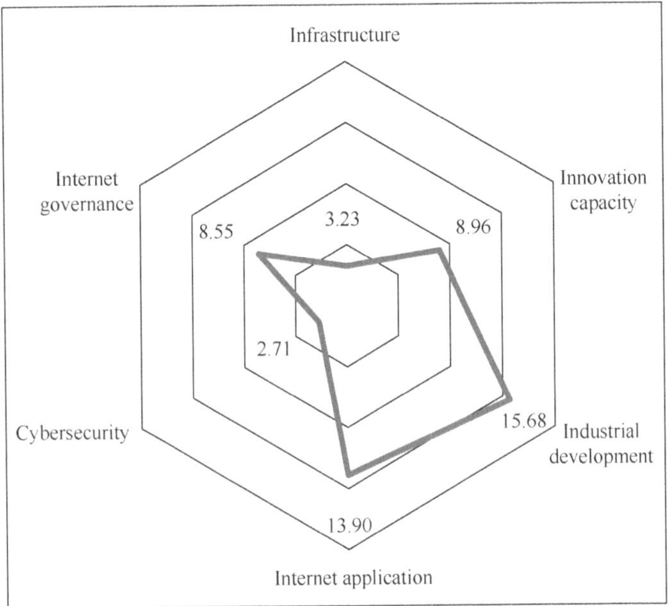

Fig. 3 Internet development index of China

patent data company, the number of patents applied by Chinese manufacturers has accounted for 34% in the world by March 2019.

Chinese government is promoting the deep integration of Internet with economy in an all-round way. E-commerce and Internet information services thrive, with new business modes and forms emerging. In 2018, China's digital economy reached 31.3 trillion yuan,[4] accounting for 34.8% of its GDP. Digital economy has become a new engine for China's economic growth.

China is endeavoring to promote the development of cybersecurity industry. Technological innovation on cybersecurity keeps active. A number of cybersecurity companies are growing rapidly. With more and more new products and services, the comprehensive strength is steadily increasing and the scale is developing continually.

In China, mobile Internet applications are more diverse, and apps closer to production and life grow rapidly. Mini programs, short video, and live streaming applications extend the users' time on mobile phones. Apps on mobile games, life services, and mobile shopping are used daily by users. It is a distinct trend that offline traffic such as fresh produce delivery and takeaway food ordering and ticketing nurtures online service, and that access traffic of mobile Internet keeps a high-speed growth. The data from App Annie shows that in 2018, global app downloads exceeded 194 billion, among which Chinese market ranked the first and accounted for around 50%.

[4]Data source: Digital China Construction Development Report 2018 issued by Cyberspace Administration of China.

3.3 South Korea

South Korea ranks the third in Internet Development Index. Specifically, infrastructure remains advanced (ranking the 8th); innovation capacity (ranking the 4th), industrial development (ranking the 5th), Internet application (ranking the 10th), and Internet governance (ranking the 10th) develop steadily; and cybersecurity (the 27th together) needs to be improved. See Fig. 4 for Internet Development Index of South Korea.

South Korea is one of the countries that own the best Internet infrastructure, with leading internet speed. As a survey of Ookla, an Internet speed testing company, suggests, South Korea's mobile Internet download speed has ranked the first by May 2019 and reached 76.74 Mb/s. In Seoul, free wireless network services are provided in both public transport and public places, such as subway, taxi, and airport. See Fig. 4 for Internet Development Index of South Korea.

South Korea enjoys developed Internet applications, especially in social media and games. As a Korean social software, KakaoTalk owns the most users and monopolizes Korean market, with a market share of 94.4%. Since South Korea built 5G commercial networks, three Korean telecom operators have handled 5G mobile phone access procedures for the public and launched several 5G-based application services. In addition to 5G network deployment, South Korea actively participates in the formulation of relevant standards.

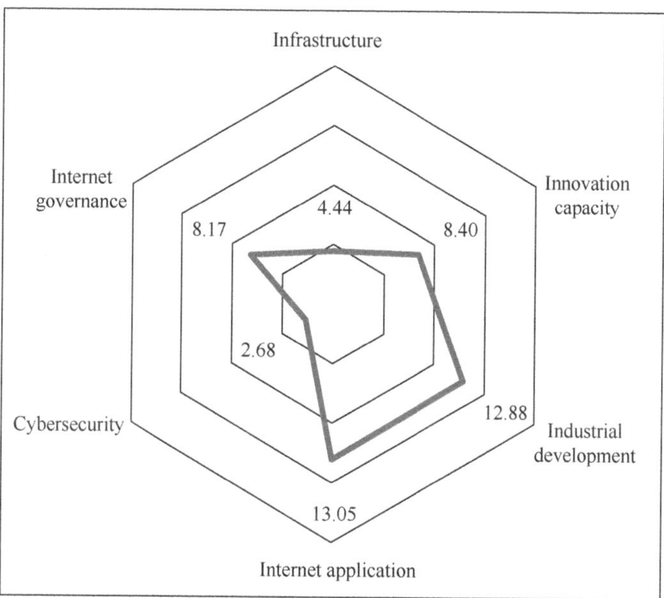

Fig. 4 Internet development index of South Korea

South Korea government keeps stipulating and improving law system of cybersecurity. It has issued National Cybersecurity Strategy and established a law system of cybersecurity that consists of relevant specialized laws and comprehensive laws on cybersecurity management, critical information infrastructure protection, information and communication networks stability safeguard, and data security. South Korea is also one of the countries in the world that first set up Internet censorship institution and legislated on Internet censorship.

3.4 France

France ranks the 5th in Internet Development Index and boasts balanced overall strength. It ranks top in infrastructure (the 6th), innovation capacity (the 8th), industrial development (the 14th), Internet application (the 9th), cybersecurity (the 5th), and Internet governance (the 2nd together). See Fig. 5 for Internet Development Index of France.

France enjoys prominent advantage in innovation information infrastructure innovation and human capital and research. In Paris, Lyon, and Grenoble, there are numerous innovation clusters with strong innovation capacity. The data of France INSEE shows that the number of newly founded enterprises increased by 17% in 2018, owing to labor market and tax reform. In addition, France has made great

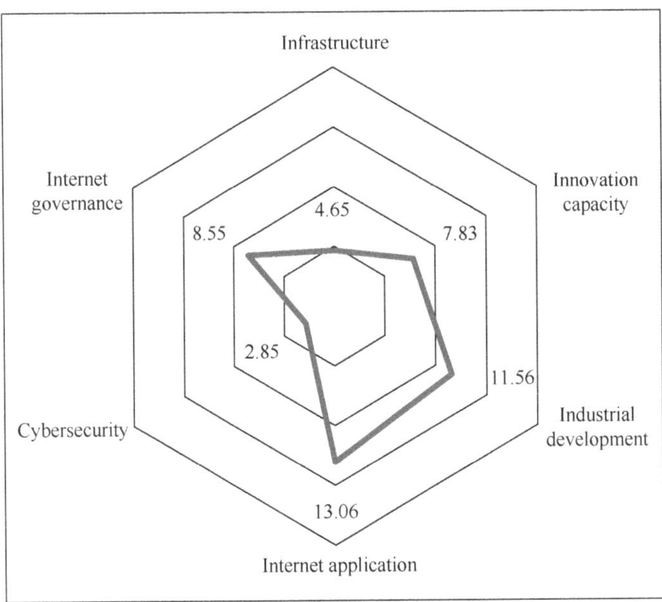

Fig. 5 Internet development index of France

breakthroughs in the field of knowledge and technological output, especially in the remarkable rise in computer software.

In industrial development, France has global well-known communication operators such as Orange, SFR, and Free Mobile, high-tech giants such as Atos and Dassault, as well as several newly founded enterprises with value of assessment over $1 billion. By 2018, in France, there were 109 newly founded enterprises in artificial intelligence, accounting for 3.1% in the world. They were distributed in such fields as health care, manufacturing, transportation, public services, environment, and financial services.

French military department is planning to develop and deploy offensive cyber-weapons to improve protective level. By 2025, France will allocate funds to hire 1000 cyber-soldiers to enhance the construction of cyber-forces. French Defense Ministry issued Offensive Military Cyber-Warfare Doctrine, to combine traditional military operations with cyber-warfare and safeguard national sovereignty and security.

3.5 Germany

Germany ranks the 9th in Internet Development Index. Specifically, it ranks the 16th in infrastructure, the 5th in innovation capacity, the 25th in industrial development, the 6th in Internet application, the 4th in cybersecurity, and the 2nd in Internet governance. See Fig. 6 for Internet Development Index of Germany.

German government has increased investment in information infrastructure. In 2018, it established "Special Fund for Digital Infrastructure" and invested € 2.4 billion to promote broadband Internet construction and "Digital School" project. In next four years, the special fund will enlarge and play greater role. In July 2019, Deutsche Telekom announced the launch of commercial 5G network, which would cover 20 cities by the end of 2020.

German government has continuously increased investment in innovation. In order to further stimulate scientific research and innovation and promote the transformation of innovative achievements, Fraunhofer Society and European Investment Fund (EIF) decided to jointly set up a fund called "Fraunhofer Tech Transformation Fund." With a fund size of € 60 million, it is to accelerate the market transformation of intellectual property rights in 72 institutes and organizations affiliated to it.

Germany strives to develop "Germany made AI" and supports universities, institutes, and companies to build international AI labs, so as to promote international cooperation on scientific research for excellent performance, advance innovative research, and strengthen the transfer of knowledge and technology.

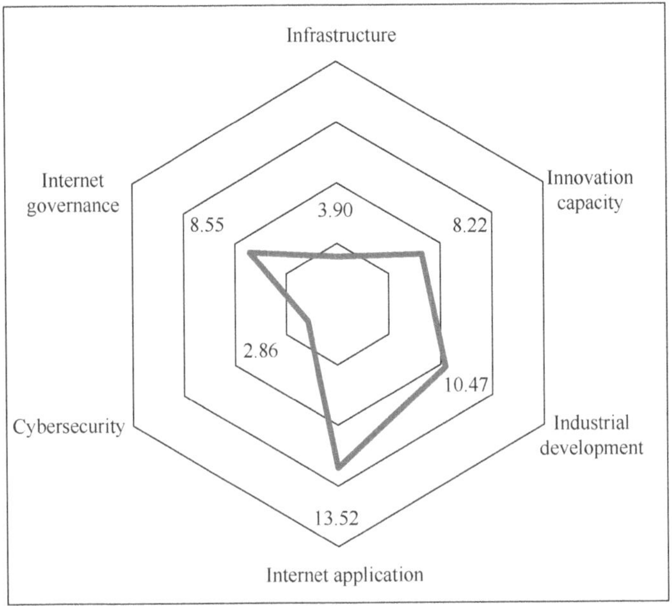

Fig. 6 Internet development index of Germany

3.6 Israel

Israel ranks the 14th in Internet Development Index. Specifically, it ranks the 15th in infrastructure, the 10th in innovation capacity, the 3rd in industrial development, the 39th in Internet application, the 2nd in cybersecurity and the 24th in Internet governance. See Fig. 7 for Internet Development Index of Israel.

Israel takes the lead in technological innovation capacity in semiconductor, virtual reality, artificial intelligence in the world. In August 2018, World Economic Forum (WEF) published the Global Competitiveness Report 2018. It revealed that in Israel, R&D investment accounted for 4.3% of its GDP, ranking the first in the world. In Herzliya, a city in the north of Israel, multinational high-tech companies assemble, giving it the name "Minor Silicon Valley."

Israel nurtures many newly founded enterprises and provides more chances for technological innovation. The Global Competitiveness Report 2018 suggests that Israel ranks the first in entrepreneurship risk tolerance and development of newly founded enterprises in the world. According to a report issued by Israel Innovation Authority, Israel owns the largest number of newly founded enterprises per capita in the world. Each year, 140 AI enterprises were newly founded averagely. Now, more than 950 active newly founded enterprises are using or developing AI technology.

Israel displays prominent advantage in cybersecurity and ranks top in the world. In Israel, there are more than 400 Internet companies and 50 R&D centers of multinational corporations. Famous international companies, such as Deutsche Telekom,

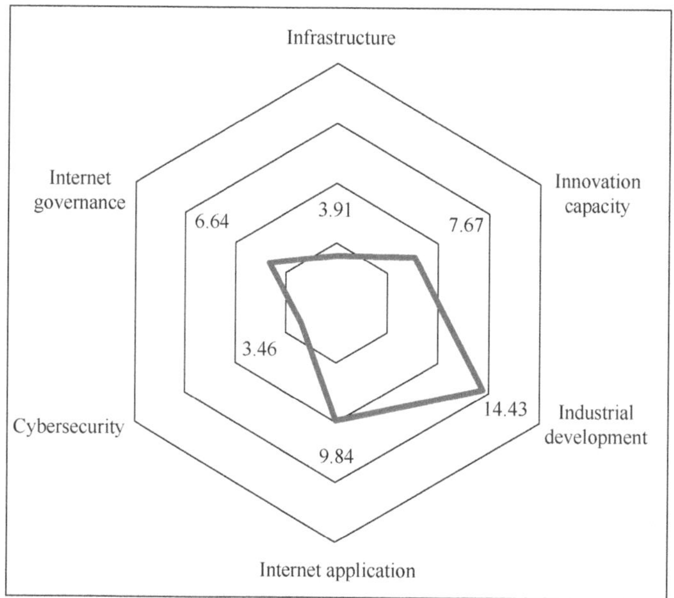

Fig. 7 Internet development index of Israel

EMC, Lockheed Martin, Citibank, PayPal, General Electric, Amazon, Cisco, Intel, AVG, and Oracle, set up cybersecurity R&D centers in Israel. Over 90% of Global Top 500 Enterprises adopted Israel's cybersecurity solution. Currently, Israel has mastered virus invasion, program destruction and cyber-deception, and can conduct targeted cyberattacks.

3.7 Russia

In 2019 Internet Development Index, Russia ranks the 17th. As the first country which developed modern Internet, artificial intelligence, and digital currency, Russia ranks the 9th in infrastructure, the 20th in innovation capacity, the 19th in industrial development, the 32th in Internet application, the 27th in cybersecurity, and the 7th in Internet governance. See Fig. 8 for Internet Development Index of Russia.

Now, Russia is accelerating commercial uses of 5G and promoting economic transformation. In particular, Moscow has laid foundation for 5G development and attracted investment from major telecom operators. Russia strives to build Moscow into the first 5G city by 2020.

In terms of innovation capacity, in Skolkovo Innovation Center in the outskirts of Moscow, as Silicon Valley of Russia, there are about 2000 high-tech companies settled, focusing on developing multimedia search engines, image recognition

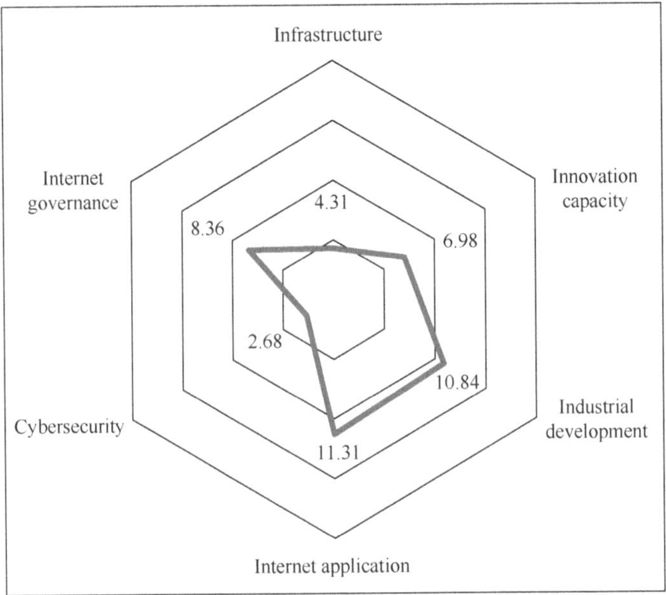

Fig. 8 Internet development index of Russia

and processing technologies, analytical software, smartphone applications, new-generation data transmission and storage, cloud computing, information security, wireless sensor network, and medical information technologies.

Aiming to improve Internet comprehensive governance, Russia government has strengthened top-level design and applied heavier punishment. In March 2019, Russian President Vladimir Putin signed Disrespect Laws and Fake News' Law, to exert severer penalties on "those who openly spread offensive information against Russian society, constitution and government" and to crack down the spread of misinformation and insults online. In May 2019, Russian President Vladimir Putin signed Russia's Sovereign Internet Law to ensure that Russia's Internet runs stably in case that external network is cut off. It is an "defense operation in advance" to counteract current problems and threats in cyberspace.

3.8 India

In Internet Development Index, India ranks the 23rd, thereinto, it ranks the 42nd in infrastructure, the 13th in innovation capacity, the 21st in industrial development, the 11th in Internet application, the 38th in cybersecurity, and the 19th in Internet governance. See Fig. 9 for Internet Development Index of India.

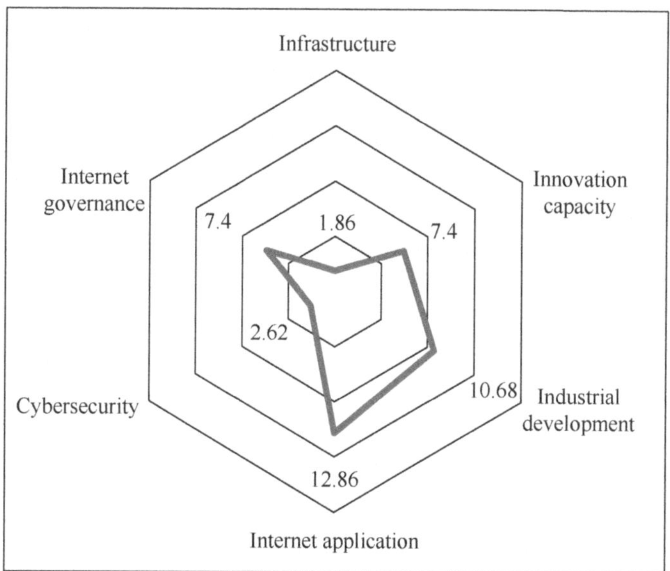

Fig. 9 Internet development index of India

In development of information infrastructure, India lags behind but it owns great potential. In recent years, India has made rapid progress. Number of Internet users and Internet penetration rate grew substantially. According to the statistics of Internet Trends Report 2019, Indian Internet users in 2018 accounted for 12% in the world, second only to China. In terms of growth of Internet users, it increased 97.89 million users, by 21% over the past year. Internet penetration rate reached around 41%, a great leap comparing with 31% in 2017. GSM Association predicted that by 2025, 50% Americans, one-third of Chinese, and only 3% Indian will be connected to 5G services. In the long term, Indian 5G market is huge.

India values R&D investment in innovation. It owns high-quality publications and colleges on science and its international competitiveness and innovation capacity have improved constantly. In July 2019, World Intellectual Property Organization (WIPO) released Global Innovation Index 2019, in which India was regarded as a pioneer in Central and South Asia, and Bangalore, Mumbai, and New Delhi were included in the list of Global Top 100 Technology Clusters.

Indian Internet venture capital (VC) is booming. With a population of 1.3 billion and mobile Internet users of 560 million, India has achieved rapid growth in Internet industry over the past five years. International investors and IT practitioners view India as the next "one-billion market" after China. According to the data published by Tracxn, a market analyst company, in March 2019, China's VC investment in India reached $5.6 billion in 2018, over 8.3 times of $ 668 million in 2016. Inc42, an Indian media platform, stated in January 2019, that Indian newly founded enterprises covering 743 fields obtained $11 billion in 2018. Among them, fintech and e-commerce received the most attention from investors.

India is the largest consumer of mobile data in the world, with the fastest-growing market for Internet and smart phones. India is now the second-largest telecommunication market in the world second only to China. India has an urban population of over 400 million and a rural population of over 900 million. In cities, there are 295 million mobile phone network users, and in countryside, there are about 200 million Internet users. What is more, the numbers are increasing. GSM Association expected that the number of Indian mobile phone network users will grow at a rate of 6% a year and reach 670 million by 2020.

3.9 Brazil

In terms of territorial area, Brazil ranks the 5th in the world and the first in South America, with a population of 212 million. In Internet Development Index, Brazil ranks the 27th. Specifically, it ranks the 41st in infrastructure, the 26th in innovation capacity, the 27th in industrial development, the 21st in Internet application, the 32nd in cybersecurity, and the 22nd in Internet governance. See Fig. 10 for Internet Development Index of Brazil.

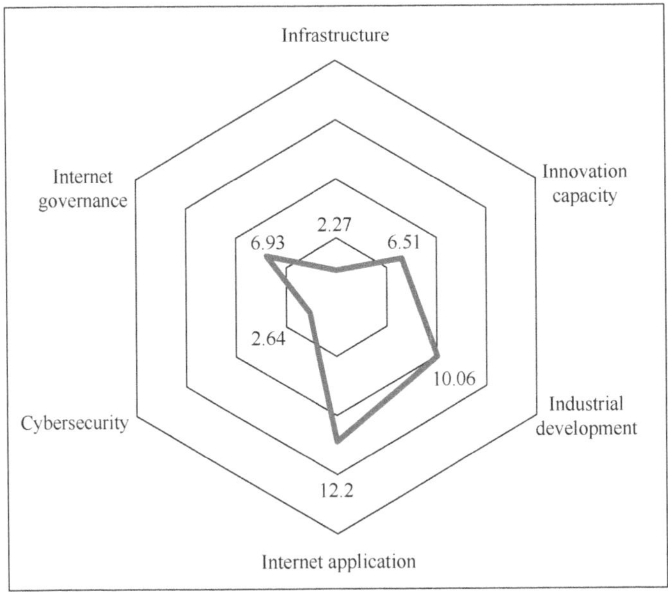

Fig. 10 Internet development index of Brazil

Brazil displays vitality in information industry and especially local newly founded enterprises. In 2018, Brazil invested $1.3 billion in venture capital ($859 million in 2017), accounting for 66% of the total investment in Latin American. As of 2018, eight newly founded enterprises in Brazil became unicorns with value over $1 billion. Brazilian information industry is developed rapidly, no matter ride-hailing, food delivery, credit services or e-banking. Innovation mainly focuses on financial system. Over 106 million Brazilians are active users of online banking, and 58% bank transactions are online.

Brazilian government has been taking much effort to encourage innovation, but it is necessary to reconsider the existing policies and make further adjustment and revision. Brazilian R&D expenditure accounts for 1.17% of its GDP, only half of the average (2.5%) of member states of Organization for Economic Cooperation and Development (OECD). In number of patent applications, Brazil accounts for only 2.2% of the total of BRICS and 0.003% of the world. The problem of talent shortage remains unsolved. About 61% of Brazilian companies admit the shortage of skilled workers. In member states of OECD, the rate is only 34%. In Brazil, 21.6% of the workforce is employed by knowledge-intensive industries, half of that of member states of OECD (39.8%). Due to tariffs and non-tariff barriers, it is hard for Brazil to introduce technologies by import. This gives less chance to Brazil and its companies to learn cutting-edge technologies in the world. Comparing with Latin American countries, Brazil maintains a higher level of innovation output. Yet, it is far lower than the average of member states of OECD. Probably, Brazil must rely on labor-intensive industries that are low-skilled and low-innovative.

Brazil is the third-largest mobile Internet user in the world. Among Internet users, mobile users account for 93%, and more than two-thirds of Brazilians own smart phones. In online time, Brazil ranks the 2nd in the world. On average, Brazilians spend 9 hours and 29 minutes online every day, much longer than global average time of 6 hours and 42 minutes. E-commerce has developed fast and Brazil became the largest e-commerce market in Latin America. According to a survey of GS&MD, a Brazilian consulting firm, the trust index of Internet users on e-commerce tools is as high as 64%, far above the world average level. Data from Brazilian government portal (brasil.gov.br) shows that e-commerce market grows by 20%-25% every year.

3.10 South Africa

In Internet Development Index, South Africa ranks the 34th. Specifically, it ranks the 40th in infrastructure, the 28th in innovation capacity, the 30th in industrial development, the 35th in Internet application, the 23rd in cybersecurity, and the 20th in Internet governance. See Fig. 11 for Internet Development Index of South Africa.

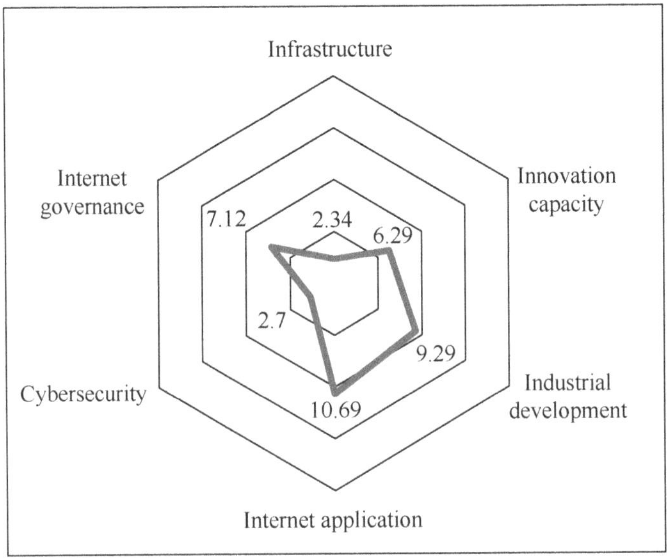

Fig. 11 Internet development index of South Africa

South Africa is committed to improving technological innovation system. In March 2019, South Africa published a new edition of White Book on scientific and technological and Innovation, placing technological innovation at the center of its development.

In industrial development, Naspers is a famous investment and Internet giant in South Africa. It is one of the top ten Internet corporations in the world and the largest shareholder of Tencent, a Chinese Internet corporation. Moreover, Naspers made investments in Allegro, Eastern Europe's largest e-commerce enterprise, Mail.ru, one of Russia's Internet giants, Flipkart, India's largest e-commerce enterprise, and several Internet assets in South America.

In Internet application, among top ten websites of global e-commerce traffic, five websites are from South Africa. In South Africa, mobile payment lags behind and logistics industry develops slowly, which affects the development of e-commerce. In social media, the top three social network platforms with most registered users are Facebook, YouTube, and Twitter. The number of registered Facebook users accounts for 20% of the total population of South Africa. See Fig. 11 for Internet Development Index of South Africa.

4 World Internet Development Prospects for 2020

In the future, world Internet development will enter a stage with significantly increasing uncertainty. As innovation accelerates and competition intensifies, non-technological elements play more important role. With 50 years of experiences, Internet development just starts, and human must wait for a long time until Internet makes subversive change to their society and life. The bursting energy of AI, IoT, 5G, IPv6, quantum computing, data technology, and blockchain relies on the integration, innovation, breakthrough and advance of technology and application that comes with Internet. With the adjustment of social division of labor in Internet environment, the change of possession forms of the means of production represented by data, and the transit of global wealth form and economic growth pattern, the impact will happen to various fields, levels, and dimensions in human economy and society, globally and gradually. This will not only test human's ability in controlling technological development but also concern the future and destiny of human.

Internet development will profoundly change and affect the future of human society. In 2020, all countries in the world should view Internet as the common home of human, strengthen the sense of a community of shared future in cyberspace, and collaborate in handling the risks and challenges in cyberspace with more positive attitude. They should enhance communication, enlarge consensus, deepen cooperation, and work together to promote discussion and co-building and sharing and co-governance in cyberspace, so that the achievements of Internet development can better benefit people in the world.

4.1 Given the Prominently Unbalanced Internet Development, It Is Imperative to Improve the Popularity of Internet

Presently, unbalanced development and level occur in Internet popularity, information infrastructure construction, technological innovation, and security risk prevention in different countries and regions. International digital divide remains unsolved, which affects and limits the informatization construction and digital transformation in the world, especially in developing countries and least developed countries. International community must undertake the historical task to promote the popularity of Internet development, formulate effective policies suitable for new technological revolution and industrial transformation, strengthen the connectivity of cyber-infrastructure, and improve the ability in getting access to Internet in rural and remote areas, so as to provide necessary support in finance, technology and talent to developing countries and least developed countries and to share the achievements of Internet development with all people in the world.

4.2 With Opportunities in the Rapid Development of Digital Economy, It Is Imperative to Promote Technological Innovation

Digital economy is a new economic form in human's social development. Now, it becomes a new growth driver for global economy and occupies an important position in global economic development. The most distinctive feature of digital economy is taking data as a critical productive factor, and it uses Internet information technology as an effective core driving force to improve total factor productivity and optimize economic structure. All countries should encourage technological innovation and creation in the field of Internet, accelerate the development of new-generation information technologies such as cloud computing, big data, blockchain, IoT and AI, advance the cultivation of new technologies, new applications and new business forms, and promote the deep integration of digital economy and traditional industries. More importantly, all countries should follow the principles of openness and cooperation for mutual benefit, accelerate technological R&D and network deployment of 5G and IPv6, and jointly boost the digital transformation of global economy.

4.3 In Face of New Challenges Confronting a Healthy Cyber-Culture, It Is Imperative to Regulate the Order in Cyberspace

Cultural diversity in cyberspace enriches human spiritual life and promotes the exchange of human civilization. Promoting the prosperity and development of cyber-culture is the common wish of people in the world. Now, the foundation for trust in cyberspace is threatened or damaged. Crimes such as Internet fraud and cyber-violence happen frequently despite of repeated bans. Some cyber-products are changing into channels of cyber-crimes, which affect and impact the normal order of cultural dissemination. All countries should work together to create an interactive platform for different cultures on Internet, to promote exchanges and mutual learning among different cultures, and form a new ecosystem of cultural integration in Internet environment. They should promote the network dissemination of excellent cultural products and carry forward the positive energy of optimism, fairness, and justice. They should make joint efforts to regulate the order in cyberspace, strengthen communication and exchange, enhance the cooperation of all parties, and help the harmonious progress of world civilization.

4.4 In the Increasingly Severe Situation, It Is Imperative to Strengthen Cybersecurity Protection

As cyberspace becomes a new key field of human values in development, cybersecurity is facing more and more severe situation. Cyberattacks turn out more complex, and hackers become more formalized, organized, industrialized, and professional, in larger scale. The attack is more frequent by new means in a larger range. Various cyberattack incidents impact global economic and social development more deeply. Global Risks Report 2018 published by World Economic Forum (WEF) states that cyberattacks, data fraud, and data stealing are listed in top five global risks in 2018. International community should strengthen mutual trust, effective cooperation, and opening and sharing to promote the formulation of global cybersecurity norms. All countries should enhance communication and negotiation in the field of cybersecurity, reinforce the protection of critical information infrastructure, raise the awareness and ability of global cybersecurity protection, and jointly build a new cybersecurity order based on security, stability and mutual trust.

4.5 As Supports for Cyber-Sovereignty Grow, It Is Imperative to Promote the Internet Governance System Reform

As more countries recognize, support, and safeguard cyber-sovereignty, the current Internet governance system fails to fit the reality. In international community, it is imperative to reform the Internet governance system. In particular, a few countries uphold unilateralism and protectionism, which disturbs global industrial chain and arouses great uncertainty to the future Internet development. International community must undertake the responsibility of promoting the Internet governance system reform. All countries should adhere to equal multilateral and multi-party participation, safeguard their rights to develop, participate, and govern in cyberspace, give full play to the role of governments, international organizations, Internet companies, tech communities, non-governmental organizations, and citizens, to form a multilateral, democratic and transparent governance system, and jointly build a community of shared future in cyberspace.

Contents

Chapter 1
Development of Information Infrastructure in the World

1.1 Outline

As the basic and strategic resources in information society, information infrastructure plays a vital role in economic and social transformation and upgrading, national security and development, and the well-being of all peoples in the world. Major countries endeavor to enhance strategic deployment, strengthen forward-looking layout, and promote the transformation and upgrading of network facilities toward new-generation information infrastructure. They made great efforts to accelerate the construction of new infrastructure such as 5G, artificial intelligence, industrial Internet, Internet of Things, and collaborated in the interconnection among global network infrastructure, which established solid foundation for seizing new opportunities in information revolution, cultivating new economic growth-drivers, recovering new national competitive advantage, and sharing new achievements of Internet development.

(1) The upgrading of broadband network facilities is accelerating. Fixed broadband has been popularized, the need of High Bandwidth such as Ultra HD sped up the deployment of gigabit optical network, and commercial uses of 5G starts. Competition on the construction of spatial information infrastructure got fierce. Especially on the competition between the launch of high-orbit high-throughput broadband satellite and low-orbit satellite constellation system got white-hot. Software Defined Network (SDN) provided stronger network controlling ability, artificial intelligence extended to telecommunication network, and telecom operators increased investment in improving the construction of intelligent network.

(2) Application infrastructure developed steadily. The penetration rate and traffic of IPv6 rose stably. Mobile network is the main driver of IPv6 deployment, and software/hardware IPv6 support degree was improved continually. New-type Internet business such as Internet of Everything (IoE), online video and VR/AR contributed to the high-speed growth of global Internet traffic. In terms of global data centers, the number was decreased and the scale was enlarged.

Chinese Academy of Cyberspace Studies, *World Internet Development Report 2019*, https://doi.org/10.1007/978-981-33-6938-2_1

The United States maintains the advantage in total number of hyperscale data centers. The need of Asia–Pacific and EMEA for data center grew rapidly. Edge computing became a focus attracting attention of the industrial circles in the world. The polymerization effect of Internet Exchange Point proved prominent, and received greater popularity worldwide. International Major Exchange Point has transformed from Network Access Point (NAP) to Internet Exchange Point (IXP) among basic operating companies.

(3) The deployment of new-type information infrastructure was accelerated. Major telecom operators sped up the deployment of Cellular Internet of Things. As the application scenarios of Internet of Things largely expanded, NB-IoT/eMTC became the most important option of IoT technology for major telecom operators, and LoRa (Low Power LAN Wireless Standards) became a typical model of private network deployment. It is a common view to develop industrial Internet, which activated the construction of industrial Internet platform and formed three accumulation areas in the United States, Europe and Asia–Pacific.

1.2 Broadband Network

1.2.1 The Need of High Bandwidth Accelerated the Deployment of Gigabit Fiber Optical Network

1.2.1.1 Ultra HD Video and VR Promoted the Upgrading of Access Capability

Ultra HD industry has taken shape, and global users of Ultra HD have exceeded over 200 million households. The popularity rate of 4 K/8 K TV is increasingly higher, with more high-quality products such as 4 K TV programs, movies, live streaming and games emerging. The improvement of video quality put forward higher requirements to network capability. In particular, Ultra HD live streaming requires stable and effective network. And the problems of low-latency and low packet loss rate must be solved. As global VR industry got advanced with in-depth application, 2D video extended to 3D video, mainly including VR. In spherical panorama 360° video and spatial entity video, the video broadband traffic must be increased by 10–100 times, which demands higher-quality bandwidth.

1.2.1.2 The Deployment of Fiber Broadband Network was Accelerated

The deployment of such technologies as 10G PON and DOCSIS 3.0 enabled the access capacity of fiber broadband to reach gigabit. In China, Shanghai strived to build "the First City with Double Gigabit" and realize the goal of full coverage in gigabit broadband. By the third quarter of 2019, Shanghai had achieved gigabit broadband coverage of 5.6 million households and 3,000 buildings. The popularity

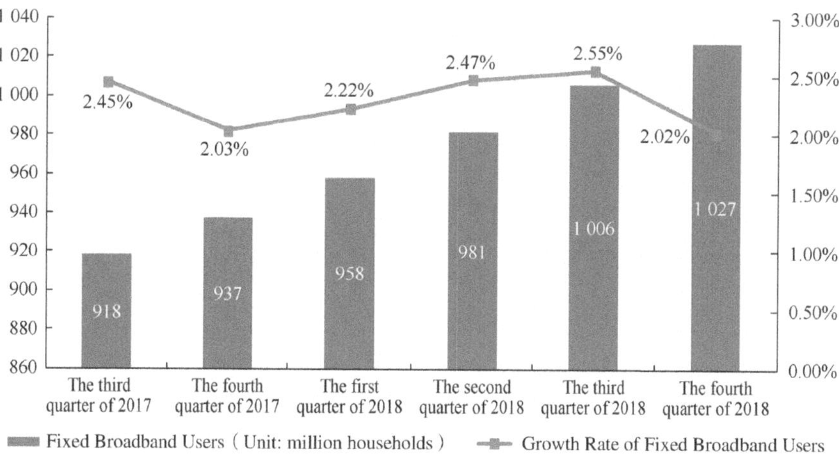

Fig. 1.1 Development of global fixed broadband users from the third quarter of 2017 to the fourth quarter of 2018. *Data Source* Point Topic

rate and users of fixed broadband continued to grow. According to Point Topic's data, by the end of 2018, there were 1.027 billion users of fixed broadband in the world, with an annual increase by 9.58%. Technologically, copper cable connections continued to decrease, by 8% year on year; fiber optic connections increased, by 22%; cable users increased, by 3%. China is still the biggest market for the growth in fiber. By the end of 2018, the connections of Fiber To The Premises (FTTP) increased by 24%, equivalent to 74% of global FTTH net growth over the same period. Other countries with remarkable growth in FTTP were Thailand, Ireland, Italy and Brazil, with the growth rate of 35%, 21%, 15% and 14%, respectively.

See Fig. 1.1 for development of global fixed broadband users from the third quarter of 2017 to the fourth quarter of 2018. See Fig. 1.2 for regional market shares of technology in global fixed broadband in the fourth quarter of 2018.

1.2.2 Mobile Broadband is Evolving Toward 5G

1.2.2.1 4G Construction Goes Stably

According to a research report of Global Mobile Suppliers Association (GSA), by June 2019, 752 operators have operated 4G LTE (Long Term Evolution) networks and provided mobile or fixed wireless access services in 223 countries and regions. Among them, at least 160 operators have launched LTE-TDD network. As the statistics of GSMA Intelligence suggests, by the end of June 2019, there were 7.84 billion mobile users in the world, with the popularity rate at 101.6%. In specific, 3.74 billion were 4G users, accounting for 47.7%.

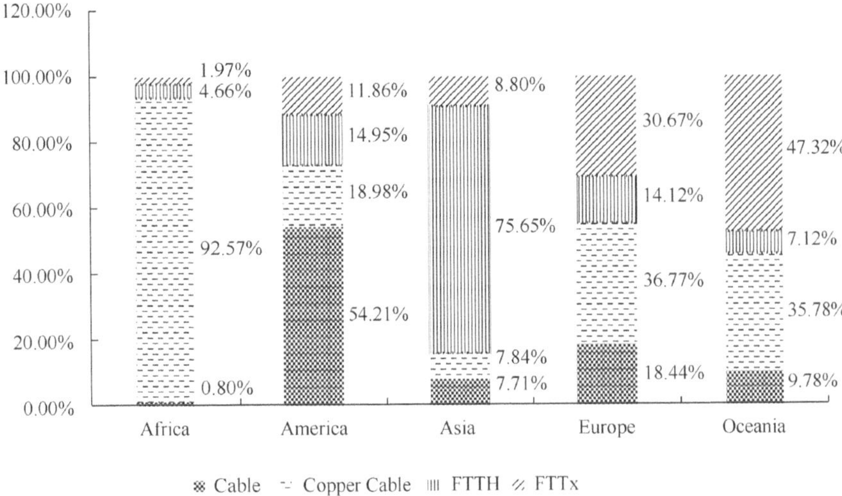

Fig. 1.2 Regional market shares of technology in global fixed broadband in the fourth quarter of 2018. *Data Source* Point Topic

See Fig. 1.3 for development of global mobile broadband users in different regions.

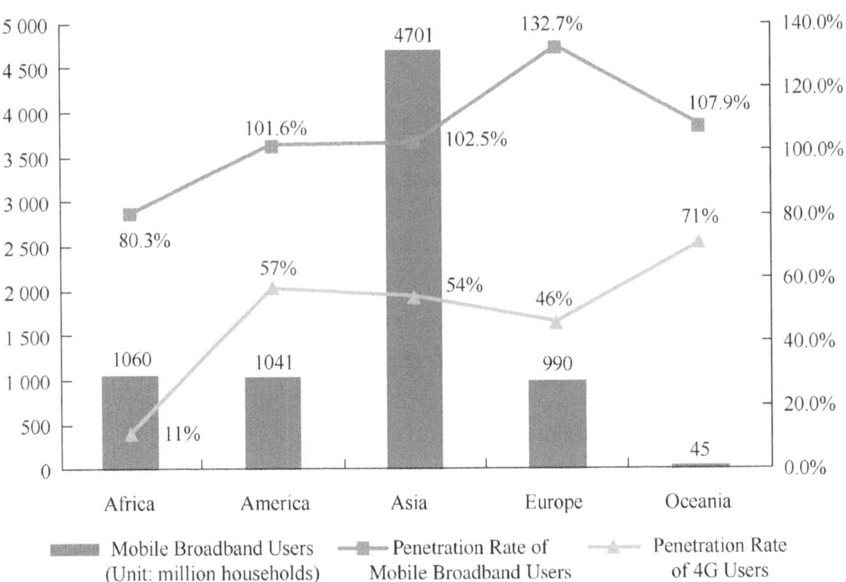

Fig. 1.3 Development of global mobile broadband users in different regions (By the end of June 2019). *Data Source* GSMA Intelligence

1.2.2.2 Commercial Uses of 5G Started

2019 is the first year for commercial uses of 5G. By the end of June 2019, there were 2.03 million 5G users in the world, and 280 operators in 94 countries invested in 5G network in the form of test, trial, pilot, plan and actual deployment. Specifically, 26 operators in 16 countries including South Korea, the United States, Switzerland, Italy, the United Kingdom, the United Arab Emirates, Spain and Kuwait, were able to provide services on commercial uses of 5G that meet 3GPP standards. On April 3, 2019, South Korea took the lead in launching services and became the first country in the world that made commercial 5G network open to ordinary users. Simultaneously, the United States started commercial uses of 5G. On June 6, 2019, China officially issued 5G licenses, and Chinese operators started to build 5G networks extensively to cover 40 cities first and distribute numbers to the public later.

See Fig. 1.4 for the percentage of 2G, 3G, 4G, and 5G users in the world in the second quarter in 2014–2019.

(1) 5G technological standards got mature. In the light of the latest timetable of 3GPP 5G Air Interface standard (Rel-15/16/17), Rel-15 has been fully completed and frozen (no new features added), Rel-16 is in progress and its freezing time was postponed, and Rel-17 is under preparation. Currently, the services on global commercial uses of 5G are still mainly based on the standard Rel-15NSA mode released on March 2019. As the standard 5G 2.0 version, Rel-16 is mainly used in vertical industries and overall system upgrading. It plays roles in developing 5G V2X (smart car and traffic), enhancing Industrial Internet of Things and Ultra Reliable and Low Latency Communication (uRLLC), and strengthening 5G NR capability that can fully replace wired Ethernet in firms, such as Time Sensitive Networking (TSN). Now, Rel-16 is

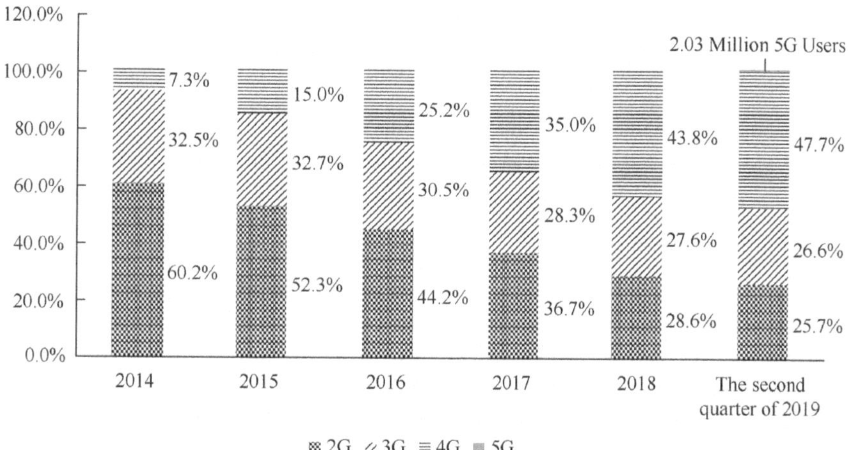

Fig. 1.4 The Percentage of 2G, 3G, 4G, and 5G users in the world in the second quarter in 2014–2019. *Data Source* GSMA Intelligence

under formulation, and its freezing time has been postponed from December 2019 to March 2020. ASN.1's freezing time has been postponed to June 2020. 3GPP will finalize the approval of Rel-17 in December 2019, formulate Rel-17 later, and freeze it in June 2021.

(2) 5G spectrums assignment focuses on various fields. In 5G age, mobile communications not only meet the communication needs among people, but also those among machines and things. Spectrum became a rare strategic resource. In face of massive connection demands, there are huge gap and great pressure. In the spectrum planning framework of International Telecommunication Union (ITU), the countries publicized their 5G spectrum strategies based on their division and use of frequency. According to a report of Global Mobile Suppliers Association (GSA), most 5G frequency bands are 700 MHz, 3400–3800 MHz and 24–29.5 GHz, and 5G deployment are mostly distributed in 3.5G MF. United States Federal Communications Commission (FCC) provided lots of HF millimeter wave frequency band for commercial 5G network deployment. EU used 700 MHz, 3.4–3.8 GHz, 24.25–27.5 GHz, 31.8–33.4 GHz and 40.5–43.5 GHz as 5G frequency bands. In Asia–Pacific, China allocated three frequency bands, 2.6 GHz, 3.5 GHz and 4.9 GHz, to 5G operations. South Korea's three major operators now are using 3.5 GHz in their commercial 5G network.

(3) Hybrid network became the favorable plan. Currently, in the world, all countries that have realized commercial uses of 5G adopted NSA-SA hybrid network plan and chose NSA to build networking faster in early stage. China's telecom operators chose different plans. China Telecom built an inter-provincial and trans-regional test network based on SA and the hybrid network of NSA-SA. China Mobile achieved the commercial use of 2.6 GHz end-to-end NSA and will realize the pre-commercial trial of 2.6 GHz end-to-end SA. Singapore IMDA supported the deployment of 5G SA to provide all 5G-related functions, such as network slicing and low-latency connection. Yet, telecom operators such as SingTel preferred to consider commercial 5G network deployment and suggest NSA networking mode.

1.2.2.3 All Countries Endeavored to Co-Build and Share 5G Supporting Facilities

To share information infrastructure means an essential foundation for the implementation of 5G. All countries actively took measures to promote the co-building and sharing of information infrastructure such as towers, open public facility resources such as buildings and tower poles to facilitate 5G construction. British Telecom (BT) worked actively with local or regional and national governments, hoping to install base station antennas on lampposts and other tall buildings. The United States aimed to remove obstacles for 5G construction at the federal, state, and local levels. United States Federal Communications Commission (FCC) has reformed some rules to fit the development of cellular mobile communications. So far, half of the states have passed the reform law on new base station location to facilitate 5G construction and reduce the cost of commercial 5G network deployment. Brazil will review pole-sharing protocol to guarantee the access to distribution pillar under fair, reasonable and non-discriminatory conditions, and to support the installation of high capacity optical fiber network and the deployment of 5G antenna.

1.2.3 Competition on the Construction of Spatial Information Infrastructure Got Fierce

Global competition on the construction of spatial information infrastructure got fierce, and especially that between the launch of high-orbit high-throughput broadband satellite and low-orbit satellite constellation system got white-hot. Global Navigation Satellite System (GNSS) has improved and provided more precise services.

1.2.3.1 Global High-Orbit High-Throughput Broadband Satellite is Developing Stably

High-orbit high-throughput satellite is developing toward broadband network, global coverage, HF communications, flexible satellite payloads, tablet terminal antenna, mobile application, diversified operation, and integration of the sky and ground. The higher frequency bands Ka (26.5–40 GHz) and V (136–174 MHz) transponders were put into use. Currently, the network bandwidth of launched high-orbit high-throughput satellite reached more than 100 Gb/s, and one satellite of ViaSat-3 (under development) will be measured with Tb/s. Horizons-3e, Intelsat's latest high-orbit high-throughput satellite, has started to serve Asia–Pacific since the first quarter of 2019. AMOS-17, Spacecom's (Israel) high-orbit high-throughput satellite, was launched in August 2019 and covered more areas in Africa, the Middle East and Europe. Kacific-1, Kacific's (Singapore) first high-orbit high-throughput satellite, was launched in the third quarter of 2019. Its small terminal can provide Internet access at more than 100 Mb/s of at low cost.

1.2.3.2 Global Competition on Low-Orbit Satellite Constellation System Got White-Hot

In recent years, middle and low-orbit satellite constellation is a hot topic in the research and investment of satellite market. Featured with all-round coverage and broadband network, middle and low-orbit satellite constellation costs less to construct and displays more flexibility, comparing with the mature high-orbit high-throughput satellite. At the end of 2018, Chinese corporations successfully launched first satellites at Hongyun constellation and Hongyan constellation, which, during on-orbit, will carry out several tests on functional verification and provide great support for the overall construction of the follow-up system. In February 2019, OneWeb (the United States) successfully launched the first six satellites, which unlocked that OneWeb's constellation plan turned from conceptual test to commercial operation. Predictably, OneWeb will build a low-earth-orbit satellite constellation that consists of 588 satellites by 2021. At the end of May 2019, Starlink of SpaceX (the United States) succeeded in launching the first 60 satellites and sent them to final orbit, and Starlink constellation network commenced.

1.2.3.3 Global Navigation Satellite System (GNSS) is Developing Steadily

At present, there are four global navigation satellite systems in the world: the United States' Global Positioning System (GPS), China's BeiDou Navigation Satellite System (BDS), Europe's Galileo Satellite Navigation System and Russia's GLONASS, and two regional navigation satellite systems: India's Indian Regional Navigation Satellite System (IRNSS) and Japan's Quasi-Zenith Satellite System (QZSS). The six systems have various advantages in military and civil affairs, stable and practical value. In recent years, they have been constructed and improved.

At the end of 2018, the United States successfully launched the first GPSIII satellite. Comparing with GPSII, GPSIII is three times in degree of accuracy and eight times in anti-jamming capability. And its lifetime was extended to 15 years. By the end of June 2019, China's BeiDou Navigation Satellite System (BDS) included 46 on-orbit satellites. With global coverage, better performance and more service, it can provide global users with distress alarming and positioning services. With around 30 on-orbit satellites, Russia's GLONASS can provide global users with navigation and positioning services on land, sea and air. Europe's GALILEO System has basically achieved global signal coverage, and as estimated, it will achieve all the satellite networks in 2020. India's IRNSS mainly provides navigation services to users in India and its surrounding areas with a radius of 1500 km. Japan's QZSS serves as a complement to the United States GPS. As planned, its satellites will increase to 7 and achieve independence and high-accuracy positioning.

1.2.4 Construction of Global Submarine Optical Fiber Cable and Terrestrial Optical Cable is Keeping Advancing

1.2.4.1 The Growth Speed of Global Internet Bandwidth Slowed Down

As of June 2019, global Internet bandwidth has reached 466 Tb/s. With a Compound Annual Growth Rate (CAGR) of around 28%, the growth slowed down. In the world, the growth speed of regional Internet bandwidth varies. In Africa, it grows the fastest, with a CAGR of 45%. Asia follows Africa with a CAGR of 42%. North America still takes the lead in inter-regional bandwidth centralization. In terms of inter-regional Internet traffic flow, the United States and Canada remain the major destinations for global Internet bandwidth connections; however, the traffic flow showed a downward trend in recent years. Relying on low IP transfer price, rich opportunities in peer-to-peer networking, favorable geographical location and abundant access to submarine optical fiber cable, Europe attracted more than 60% traffic flows in the Middle East, North Africa and Sub-Saharan Africa. See Fig. 1.5 for changes of regional traffic flows to North America in 2010–2019 and Fig. 1.6 for changes of European sub-regional capacity in 2010–2019.

Content Provider (CP) drives the growth of global Internet traffic. According to a research report of Sandvine, Netflix contributed 15% of global broadband downstream traffic, games became major contributors to traffic, live streaming began to make obvious influences on network, and the World Cup and Super Bowl are contributors to global traffic peak times, exceeding YouTube and other video applications.

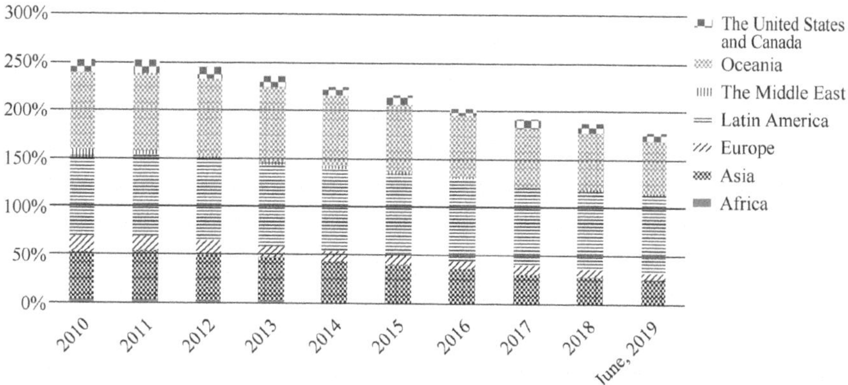

Fig. 1.5 Changes of regional traffic flows to North America in 2010–2019. *Data Source* Telegeography

Fig. 1.6 Changes of European sub-regional capacity in 2010–2019. *Data Source* Telegeography

1.2.4.2 Submarine Optical Fiber Cable is Keeping Extending

As the central nerve system of global Internet, global submarine optical fiber cable is still the key field that receives global attention and investment. According to the statistics of China Academy of Information and Communications Technology (CAICT), since June 2018, a number of global submarine optical fiber cables, such as JUPITER, SAIL, BRUSA, Hawaiki and ASC, have been completed and put into operation. Meanwhile, more than 10 global submarine optical fiber cable systems are under construction or to be started. According to the data TeleGeography releases, the number of global submarine optical fiber cable has reached 448, with a total length of more than 1.2 million kilometers. As the access directions of newly-built international submarine optical fiber cable suggest, the United States remains the main destination, and Europe and the Asia–Pacific take positive attitude toward the construction of global submarine optical fiber cables, with the United States, Japan, Singapore, the United Kingdom, Brazil and China as major participants of international submarine optical fiber cable construction. As new-emerging force of the construction, Internet giants such as Google, Microsoft and Facebook fully took part in the construction of global submarine optical fiber cable in North America-Europe, North America-Asia, North America-South America, and Africa-Europe, so as to meet the interconnection needs of their global Internet data centers. At present, there are more than 20 international submarine optical fiber cables for which Internet media companies invested or take the lead in the construction.

1.2.4.3 United Nations Actively Promoted the Construction of Asia–Pacific Information Superhighway

United Nations Economic and Social Commission for Asia and the Pacific (ESCAP) takes active and overall steps in Asia–Pacific Information Superhighway (AP-IS) project and attaches great attention to the construction of transnational terrestrial optical fiber cable systems, by means of organizing research, promoting cross-department cooperation, and exploring the loan projects from development bank. ESCAP actively negotiated with financial institutions such as Asian Infrastructure Investment Bank (AIIB) and explored the way to establish special funds in order to speed up the construction of Asia-Europe Information Superhighway Project such as TASIM.

1.2.5 Information Network is Evolving Towards Intelligence

Artificial intelligence extended to telecommunication network. Telecom operators actively explored and practiced efficient and intelligent analytical means and technologies, including smart pipe, big data, Software Defined Network (SDN), in improving the construction of intelligent network. In supporting and maintaining network operation, major telecom operators in the world explored 5G Microcell and Microcell, wireless coverage, automatic capacity optimization, error prediction and troubleshooting, as well as intelligent arrangement and management of resources, in order to change the traditional manual maintenance mode, reduce operating cost, promote the efficiency and convenience of network operation and maintenance, and improve the accuracy of business and resource arrangement. In enlarging network-related business, telecom operators actively expanded their service capabilities and channels for vertical industries. For instance, American Telephone and Telegraph Company (AT&T), Deutsche Telekom, and Vodafone provided services of unmanned aerial vehicle, consulting service and intelligent family to help realize the digital transformation of comprehensive information services.

1.2.6 All Countries Enhanced Support for Broadband Construction in the Rural Areas

1.2.6.1 All Countries Give More Supports to the Deployment of High-Speed Broadband Network in Rural Areas

Many governments place emphasis on the construction of broadband network in the rural areas. They play an active role in improving and enriching supporting means and increasing investment. EU has launched a broadband map portal to assess the progress of Internet projects. China has strengthened its deployment of high-speed broadband network in rural areas. By the end of June 2019, the proportion of access to optical fiber in China's administrative villages reached 98%, and 100% in administrative villages in Beijing, Tianjin, Shanghai and Chongqing (municipality directly under the central government) and Jiangsu, Zhejiang, Anhui, Shandong, Henan, Guangdong and Yunnan (provinces). In the United Kingdom, National Infrastructure Productivity Fund has released and subsidized Rural Gigabit Connected All Optical Fiber Programme to ensure that rural areas keep pace with the deployment of all-optical-fiber broadband. United States Federal Communications Commission (FCC) has approved a new fund on universal telecommunication service to help 106,000 rural households and small enterprises accelerate broadband access. In addition, FCC has introduced new broadband cost model and added extra subsidy to support telecom operators to increase the minimum broadband speed of these households from 10 Mb/s (upstream)/1 Mb/s (downstream) to 25 Mb/s (upstream)/3 Mb/s (downstream). Spanish Council of Ministers has approved a fund of €150 million to extend Fiber To The Premises (FTTP) to rural areas and to help provide optical fiber access to all regions and 95% of the population by 2021.

1.2.6.2 The Countries Took Various Technological Means to Promote the Extension and Coverage of Broadband Networks in the Rural Areas

4G network coverage and Fiber To The Premises (FTTP) remain the key field in the construction of broadband network in the rural areas. Internet companies explored new technologies such as low-orbit satellite to help enlarge rural broadband network coverage. China Telecom Universal Service Pilots changed the supports direction from fiber-optic broadband network access into 4G broadband network coverage. Argentina has issued 50 MHz licenses in 11 rural areas, as a part in Promoting National Rural Area Mobile Network Coverage Program. The United States SpaceX has successfully launched the first 60 satellites in its Satellite Internet project Starlink.

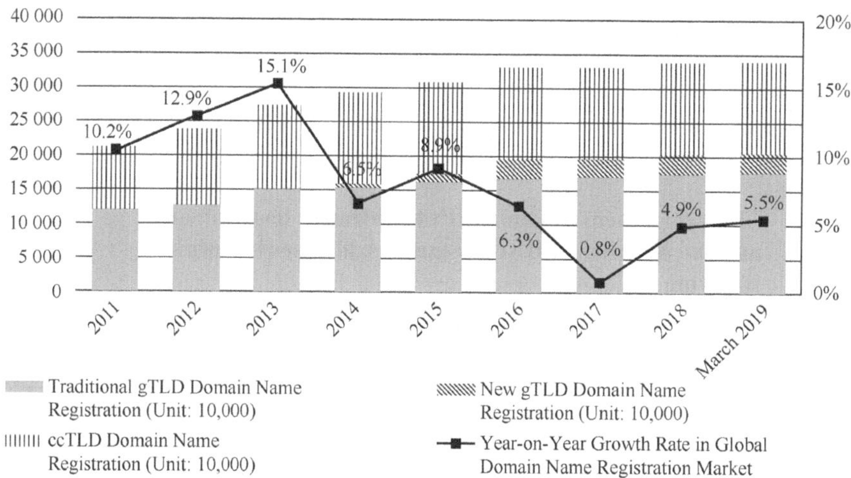

Fig. 1.7 Global domain name registration and growth from 2011 to March, 2019. *Data Source* Internet Domain Name Industry Quarterly, Ministry of Industry and Information Technology of China

1.3 Application Facilities

1.3.1 Both Domain Name Market and Facility Construction Are Growing

1.3.1.1 New gTLD Market Scale Keeps Rising and the Development of Global Domain Name Market is Accelerated

As of March 2019, global domain name registration market scale reached around 361 million, with an increase by 5.5% over March 2018 and by 0.9% over the end of 2018. In specific, ccTLD registration market scale was around 157 million, with a year-on-year growth of 7.2%, and gTLD registration market scale was around 205 million, with a year-on-year growth of 4.2%. New gTLD market scale kept rising after a fall and reached 26.869 million, with a year-on-year growth of 15.1% and accounting for 7.4% of the global domain name registration market. See Fig. 1.7 for global domain name registration and growth from 2011 to March 2019.

1.3.1.2 Number of Global Root Servers Exceeded 1000 and Domain Name Resolution Infrastructure was Keeping Improving

Root extension remains the common way to improve the performance of Domain Name System (DNS), and the number of roots continues to grow in the world. By the end of March 2019, the number of global root servers and their mirror servers had reached 1120, covering more than 140 countries and regions and providing global users with the nearest root resolution services. With the vigorous development of Internet business, most of the root server operators form global distribution patterns by setting up mirror servers to improve the resolution and security of root servers.

1.3.1.3 The Assignment of Global IPv4 Addresses has Finished and the Number of IPv6 Assigned Addresses is Keeping Growing

As of June 2019, about 3.675 billion IPv4 addresses had been assigned globally, with a notice rate of 77.23% and basically equal to that in 2018. The United States owned about 1.606 billion IPv4 addresses, accounting for 43.7% of the total number of global assigned IPv4 addresses and ranking the first in the world. China and Japan ranked the second and the third respectively. As of June 2019, the total number of global assigned IPv6 addresses had been approximately 281,348 blocks/32 (network number in 32 bits), with a year-on-year growth of around 16%. China and the United States ranked the first and the second in the total number of assigned IPv6 addresses. Yet, their usage rate was low, only with a global notice rate of around 17%. See Fig. 1.8 for top 10 countries with assigned IPv6 addresses and their notice rate.

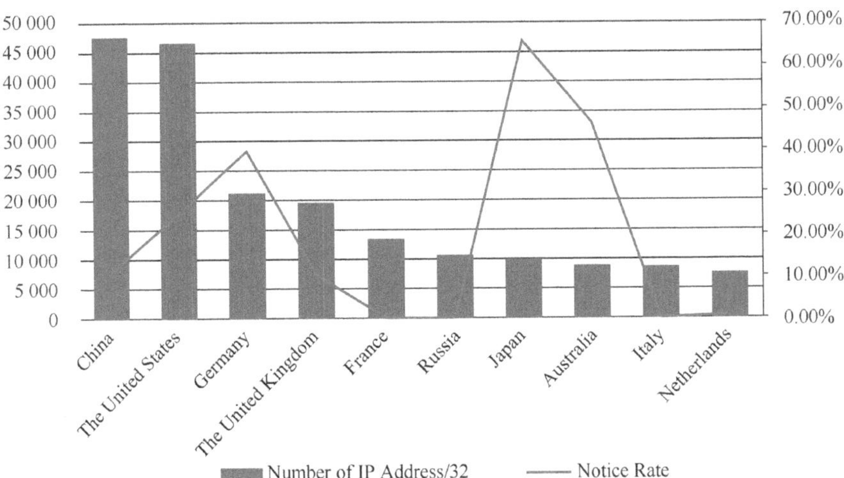

Fig. 1.8 Top 10 countries with IPv6 assigned addresses and their notice rate. *Data Source* resources.potaroo.net

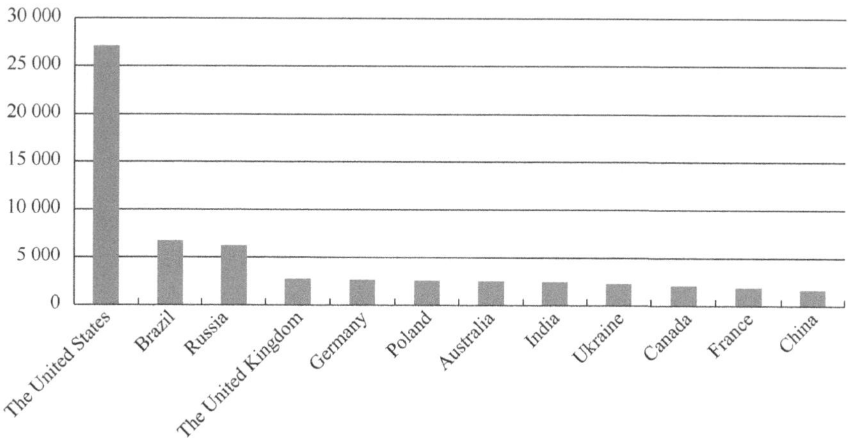

Fig. 1.9 Global applied AS numbers (Only top 12 listed). *Data Source* resources.potaroo.net

1.3.1.4 The United States Takes the Lead in the Autonomous System Number Assignment but the Global Notice Rate Needs to be Improved

As of the end of June 2019, the total applied autonomous systems (AS) number had reached 91,419 in the world, with a year-on-year growth of 4.6% and an AS notice rate of over 70%, basically equal to that in 2018. The United States owned 27,091 AS numbers and accounts for 29.63% of that all over the world. It ranked the first and keeps far ahead. Brazil and Russia ranked the second and the third respectively. See Fig. 1.9 for global applied AS numbers.

1.3.2 Commercial IPv6 Deployment is Carried Out Extensively

1.3.2.1 The Countries Actively Carry Out the IPv6 Deployment and the Popularity Rate and Traffic Rose Stably

IPv6 deployment and transfer maintained rapid development. According to the statistical data from Asia Pacific Network Information Centre (APNIC), as of the end of June 2019, global IPv6 deployment rate had reached 22.84%, with a year-on-year grow rate of 19.39%, ten times higher than 2013. Specifically, IPv6 deployment rated in North America, Asia, Europe, Oceania and South America are 31.04%, 25.27%, 17.69%, 17.93% and 1.7% respectively. In the world, IPv6 deployment is carried out in more than 170 countries, and IPv6 deployment rates are more than 5% in 62 countries and more than 15% in 36 countries. In India, the United States and Belgium,

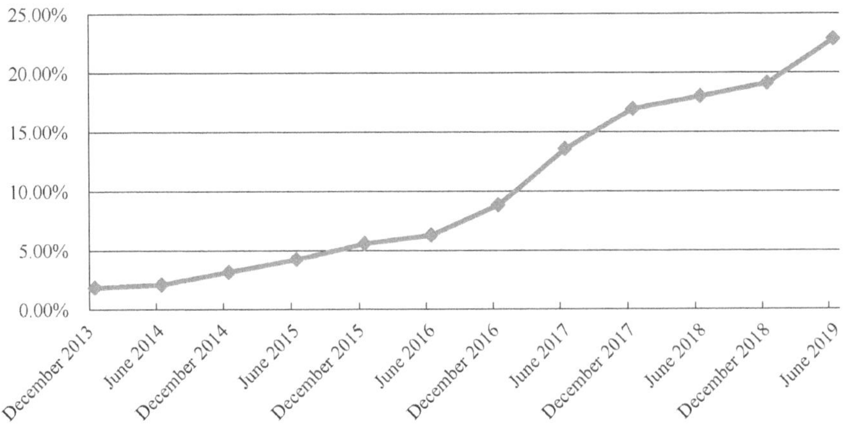

Fig. 1.10 Global IPv6 deployment rate trend. *Data Source* APNIC

it exceeded 50%. In IPv6 traffic, IPv6 traffic increased steadily in the recent year. According to the statistics of AMS-IX, IPv6 peak traffic has reached 146.6 Gb/s, with an average traffic of 116.3 Gb/s and a year-on-year growth of more than 50%. See Fig. 1.10 for the trend of global IPv6 deployment rate.

1.3.2.2 Commercial IPv6 Deployment in Global Top 10 Telecom Operators is Keeping Advancing and IPv6 Deployment Rate in Some Operators Rose

According to statistics of World IPv6 Launch, as of the end of June 2019, among global BGP routing libraries, 17,119 AS had claimed IPv6 prefixes, accounting for 26.32% of the total claimed 65,036 AS. In top 10 global telecom operators, the average IPv6 deployment rate had reached 68.94%, with a year-on-year grow of 8.33%. It exceeded 90% in T-Mobile (the United States) and Reliance Jio (India). Mobile network plays a leading role in promoting IPv6 deployment. In Reliance Jio (India) and Verizon Wireless (the United States), the traffic of mobile network exceeded 90%. See Fig. 1.11 for IPv6 deployment rate of global top 10 telecom operators.

1.3.2.3 Degree of Software/Hardware Support to IPv6 is Improving Continually and the Transformation of Website Information Source is Advancing Steadily

As the test results of Global IPv6 Testing Center suggest, by the end of April 2019, 88% of operating systems (OS) had installed IPv6 protocol stack by default, with

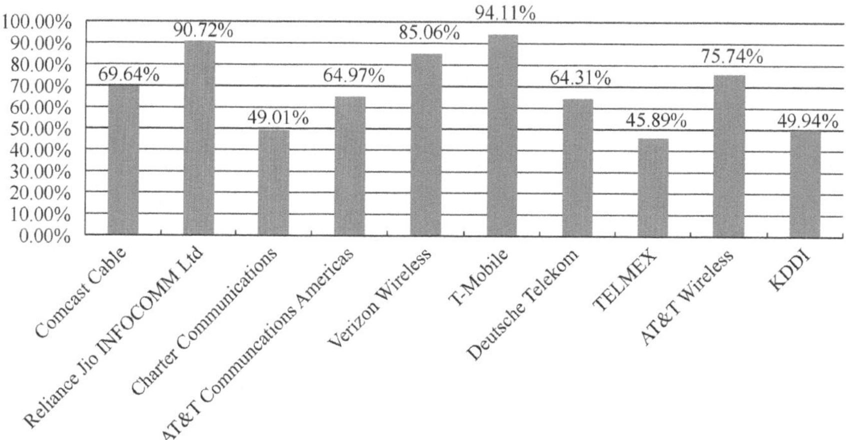

Fig. 1.11 IPv6 deployment rate of global top 10 telecom operators. *Data Source* World IPv6 Launch

70% supporting DHCPv6 and 53% supporting ND RNDSS. Main products of application software support IPv6 and network devices basically meet the demands for commercial deployment. According to the statistical data of IPv6 Ready Logo, by the end of April 2019, 2,656 IPv6 Enabled logo certifications have been issued all over the world and the number is growing stably.

1.3.3 Global Data Centers Are Expanding Rapidly

1.3.3.1 Global Data Centers are Reducing in Quantity and Increasing in Capacity Due to the Large-Scale and Intensive Development

According to a report of Gartner, global data centers began to reduce in quantity due to the large-scale and intensive development. The capacity of single frame is improving rapidly, and the quantity of racks is increasing slightly. In 2018, the quantity of global data centers reduced to about 436,000. Predictably, in 2020, it will reduce to 422,000. In the scale of rack deployment, it reached 4.899 million racks by the end of 2018, and will exceed 4.98 million in 2020, with over 62 million servers, as estimated. See Fig. 1.12 for quantity of global data centers and racks in 2015–2020.

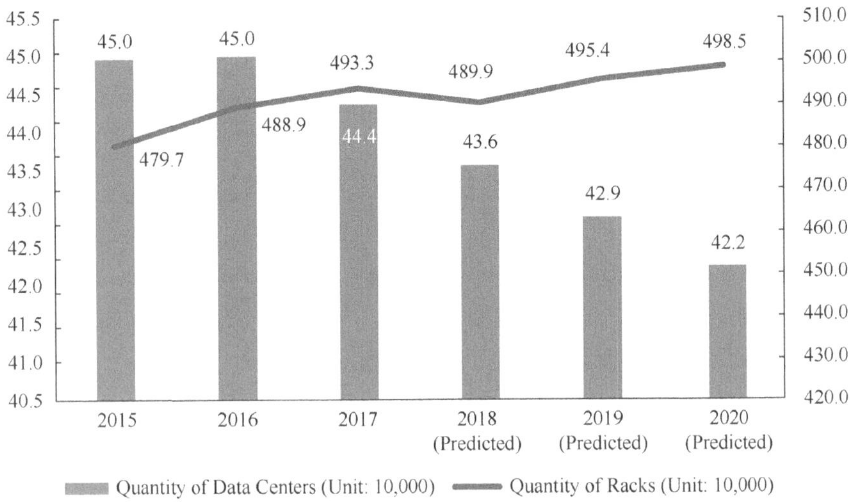

Fig. 1.12 Quantity of global data centers and racks in 2015–2020. *Data Source* Gartner

1.3.3.2 The United States Displays Remarkable Strength in Very-Large-Scale Data Centers and It's Growing Rapidly in the Asia–Pacific Area

According to a research of Synergy Research Group, in 2018, there were 430 very-large-scale data centers in the world, with an increase of 11% over 2017. In detail, the United States ranked the first in very-large-scale data center in three consecutive years, yet its share declined from 45% (2016) to 44% (2017) and 40% (2018). China, Japan, the United Kingdom, Australia and Germany shared 30% totally. Regionally, in the Asia–Pacific region Asia–Pacific region and the EMEA, the demands growed faster, with the most new-built very-large-scale data centers. Currently, the growth of very-large-scale data centers is continuing, and at least 132 very-large-scale data centers are under construction.

1.3.3.3 The Integration, M&A and Resource Expansion of Global Data Centers Stay Active

According to a survey report of Synergy Research Group, in 2018, the acquisition order number of global data center made a record, with 68 acquisition orders and transaction of $16 billion. Europe became a central area of integration and M&A. Besides, there were more cases in which resource was expanded by building new data centers, especially in the Asia–Pacific region Asia–Pacific region and EMEA. Global Switch and Equinix, two professional data center providers, invested in building new data centers in Frankfurt and London, respectively. Internet giants, Amazon Cloud Services, Google and Facebook planned to build or expand data centers in Ireland, Denmark, Sweden and Singapore, respectively, each with an investment of over $1 billion.

1.3.4 Cloud Computing and Edge Computing Are Developing in a Coordinated Way

1.3.4.1 Global Cloud Computing Giants Formally Entered Cloud Market in Africa

Regarded as an "IT wasteland", Africa owns great potential for cloud computing. Competition among cloud computing giants has extended to Africa. In 2017, Microsoft announced African Data Center Construction Plan. In March 2019, Microsoft's data centers in Johannesburg and Cape Town were officially opened, and Microsoft became the first cloud provider of cloud services relying on African data centers. In October 2018, Amazon launched Cape Town Data Center Construction Plan, which will run in the first half of 2020. Huawei will also set up two data centers in Johannesburg and Cape Town, South Africa. The construction of Johannesburg Data Center was started in 2019. It provides more low-latency and reliable cloud services to South Africa and neighboring countries.

1.3.4.2 Edge Computing Received Wide Attention from Internet Industry

Now, global Internet industry turned their attention from cloud computing to edge computing. In terms of Internet companies, Microsoft has released new-generation edge computing tools and shifted its focus from Windows operating systems to intelligent edge computing. AWS Greengrass, Amazon's edge computing platform, has revised in the form of machine learning and reasoning support. Google has announced a full shift to edge computing at Google I/O conference. Some equipment giants, such as Cisco, Huawei, and Intel, all have carried out the deployment in edge computing. With respect of telecom operators, Deutsche Telekom plans to apply edge computing to improve 5G network performance. China Mobile carried out trials of Multi-Access Edge Computing (MEC), and China Telecom collaborated with CDN enterprises on the deployment of Edge Computing Content Delivery Network (CDN). EdgeConneX, an edge data center operator, established edge data centers in Atlanta, Denver, Miami, Portland, Toronto, and Munich.

1.3.5 *Global CDN Industry is Developing Steadily*

1.3.5.1 Global CDN Traffic is Growing at High-Speed Yearly and North America Occupied the Major Market

According to statistics, as of 2018, global CDN market reached a scale of $ 9 billion, and North America has long accounted for 65% of global CDN market. In Asia–Pacific, many emerging economies gathered, with the market scale surpassing Europe and following North America. It is predicted that the ratio of global traffic carried by CDN will rise from 56% in 2017 to 72% in 2022 and CAGR in global CDN traffic will reach 44%.

1.3.5.2 CDN Leading Corporations Take the Lead and CDN Giants Focus on Cybersecurity

Akamai, an American CDN corporation, has been a global leading provider in CDN service for a long time. In 2017, it realized a revenue of $2.5 billion, accounting for 33.8% of the global CDN market. Amazon CloudFront, EdgeCast, CloudFlare and MaxCDN ranked second. As 5G age approaches, Network boundary is keeping expanding, and terminals raise higher requirement for computing and response speed, and technological iterate and innovative development of CDN, edge computing and artificial intelligence became a trend. Meanwhile, CDN became a natural network security platform with its distributed architecture, and the development of corporations turn to CDN-based edge cloud security services. Through purchasing cybersecurity companies, Akamai entered the field of cloud security and helped companies to improve the security in cloud, website, and mobile App. Wangsu Science and Technology Co., Ltd carried out the deployment in integration of edge cloud security and artificial intelligence and launched “Wangsu Shiel” Cloud Security Protection System. Based on tremendous amount of CDN security nodes, it meets the upgrading needs for protection by intelligent dispatch mechanism.

1.3.6 *Internet Exchange Point is Developing Fast*

1.3.6.1 Internet Exchange Point is Developing Rapidly and Became Global Critical Internet Infrastructure

The polymerization effect of Internet Exchange Point is prominent, and received greater popularity in the world. Presently, there are more than 900 Internet Exchange Points in 119 countries, expanding to Latin America, Africa and the Asia–Pacific region. Internet Exchange Points and its access members are increasing with a growth rate of more than 20%. In exchange traffic, the annual growth rate of "traffic" in some large-scale Internet Exchange Points exceeds 50%. In the world, there are 14 Internet Exchange Points with carried traffic measured by Tb/s.

1.3.6.2 The Access Party and Business of Internet Exchange Points have Diversified

Nowadays, International Major Exchange Points have transformed from Network Access Point (NAP) of basic operating companies to Internet Exchange Point (IXP) including Content Provider (CP), data center, Content Delivery Network (CDN), cloud service provider, and domain name/Internet monitoring infrastructure. In terms of access party, Internet Exchange Points attract lots of network access, and large Internet Exchange Points own nearly 1000 access members. In addition, as business of Internet Exchange Point got mature, the polymerization effect of network was revealed, and Internet Exchange Points no longer take basic traffic exchange as their only business. In favorable Internet ecosystem, partners gather in Internet Exchange Points and nurture innovative business involving cloud computing, cybersecurity, and IPv6. With the advantages as network hubs, International Major Exchange Points, such as AMS-IX, Equinix-IX, LINX and DE-CIX, provide testing environment for technological and business innovation, and conduct value-added innovation business, such as cloudlink, DDoS attack defense, IPv4/IPv6 transfer and hybrid cloud.

1.4 New Facilities

1.4.1 Deployment of IoT Facilities is Speeding Up

1.4.1.1 Three Leading Types of IoT Technology Providers

(1) In IoT technology, most global telecom operators chose NB-IoT/eMTC. Based on 4G network, NB-IoT provides excellent network coverage for low-medium internet speed IoT and dominates the public network. eMTC follows closely. Driven by 3GPP, NB-IoT/eMTC and their LTE technology have been included in 5G standard, which effectively ensured the smooth upgrading of NB-IoT/eMTC to future 5G networks.

(2) LoRa became a typical model in private network deployment. According to the data LoRa Alliance released in June 2019, LoRa has founded 117 network operators in 56 countries around the world. Yet, most of them are not major local operators, but urban-level or small-scale private networks.

(3) Sigfox's application scenarios are limited. As it adopts Ultra-Narrow Band (UNB), it fails to receive support of major operators and dominate the public network.

1.4.1.2 Global Major Operators Accelerate the Deployment of Cellular IoT Facilities

According to GSA's data, by the end of June 2019, 135 Cellular IoT networks have been deployed or launched all over the world. In specific, there were 98 NB-IoT networks and 37 eMTC (LTE-M) networks. Vodafone owns the most extensive NB-IoT network coverage in the world and opened commercial network in 10 countries. In its capital expenditure plan, Vodafone arranged priority to NB-IoT and planned to double the number of NB-IoT base stations in Europe before the end of 2019. China Telecom boasts the first operator that has built the most extensive coverage of commercial NB-IoT network. By May 2019, China Telecom has built more than 400,000 NB-IoT base stations and realized the full coverage in urban and rural areas. Its IoT users have exceeded 80 million.

1.4.1.3 Global IoT Application Scenarios Were Extended to More Fields

According to the latest research of Berg Insight, promoted by vigorous growth of Chinese market, the number of global Cellular IoT users reached 1.2 billion in 2018, with an increase by 70%. The installation of facilities in China accounted for 63% in the world. Meanwhile, IoT application scenarios were extended to more fields. Many innovative application programs were made in intelligent city, smart home, intelligent factory, intelligent agriculture, Internet of Vehicles and personal informatization. The penetration rate of IoT technologies and programs into various industries is continuously accelerating. It is estimated that more than 65% enterprises and organizations will apply IoT products and programs by 2020.

1.4.2 Construction of Industrial Internet Platform is Flourishing

1.4.2.1 Many Countries Made Supporting Policies on Industrial Internet

It is a common view to develop industrial Internet in the world. More and more countries formulated policies to promote the development of industrial Internet. Germany was one of the first countries that proposed the idea of industrial Internet. In February 2019, German Ministry of Economy and Energy issued *National Industrial Strategy*

2030 (Draft), which supported key industrial fields. In May 2019, Vietnamese government and Ericsson signed a series of Strategic Agreements on Promoting Industry 4.0 to speed up the development of Industry 4.0 in Vietnam. In June 2019, Japan released *2019 Manufacturing White Book* in which Japanese government proposed measures to promote the development of basic technologies in manufacturing. UK government issued a policy white paper on *Regulation for the Fourth Industrial Revolution* and outlined UK's plan to maintain its world's leading regulatory system and realize the potential in the Fourth Industrial Revolution in the period of rapid technological reform. South Korea government issued *Blueprint for the Revival of Manufacturing*, urging its manufacturing to get rid of the "quantity and late-mover" industrial mode and develop South Korea into an "innovation-led manufacturing power".

1.4.2.2 The Innovation in Global Industrial Internet Platform Maintains Dynamic

Global industrial Internet platform market keeps growing fast. In 2018, it achieved a scale of $3.27 billion, with a year-on-year growth of 27.2%. Three accumulation zones on industrial Internet platform were built in the United States, Europe and the Asia–Pacific. The United States plays a dominant role in industrial Internet platform market, and its corporations such as GE, PTC, Cisco, Rockwell and Microsoft keep driving innovation on platform edge technologies. Europe is a competitor of the United States. Corporations such as Siemens, ABB and SAP rely on manufacturing advantages and keep increasing investment in industrial Internet platform. In the Asia–Pacific, the demand for industrial Internet platform keeps strong. Asia–Pacific region represented by emerging economies such as China and India keep promoting the development of industrial Internet platform. Japanese corporations, such as Mitsubishi, Hitachi, Toshiba and NEC keep exploring new modes in industrial Internet platform and made marked achievements. The Asia–Pacific market is growing fastest and hopefully to be the largest market.

In recent years, information and communication technology (ICT) developed rapidly. Technological innovation in 5G, IoT, IPv6, cloud computing, big data, and AI accelerated and made breakthrough. They deeply integrated with economic and social fields, from Internet of Everyone to Internet of Everything (IoE), from massive data to AI, from life and consumption to production and manufacturing. They revitalized the economic and social development and took human society into a new stage. New-generation ICT plays a pivotal role in a new wave of scientific and technological revolution and industrial transformation. As it changed from partial innovation to systematic innovation, integration across industries and vertical integration became the main mode of technological innovation and industrial development. All countries will accelerate the development and application of New-generation ICT in information infrastructure and promote accelerating the digital networking and intelligent transformation of information infrastructure in order to achieve coordinated development.

Chapter 2
Development of Network Information Technology in the World

2.1 Outline

At present, global scientific and technological innovation has displayed unparalleled vitality. New-generation information technologies represented by artificial intelligence, quantum information, mobile communications, Internet of Things and blockchain, accelerated the breakthrough in the application limitation. Advanced manufacturing technologies integrating robots, digitization and new materials accelerated the promotion of intelligent, service- and green-oriented transformation of manufacturing industry. A new round of scientific and technological revolution and industrial transformation is rebuilding global innovation pattern and economic structure. In 2019, global network information technology maintained rapid development, and new technologies and new applications are developing rapidly, with emerging new achievements and breakthroughs.

(1) Basic network information technologies and cutting-edge technologies are upgraded and evolute rapidly. 7 nm Process general chips are officially used in commerce, and multi-operating systems are getting improved, with great breakthrough in hybrid brain-like chips. Advanced computing and artificial intelligence present the trend of integrate, and the open source idea exerts expanding influence. As 5G received more popularity, edge computing and virtual reality developed rapidly.
(2) Technological innovation and industrial development are daily increasingly integrated. The development of network information technology has gradually adjusted the focus, from one-field performance upgrading to integrated development of all-round industry. The full set of tools and value-added services have become the mainstream products of information technologies.
(3) Major countries play a more active role in innovation. Network information technology became a scientific and technological innovation field that attracted most R&D investment and most innovative staff, produced mostly widely used technologies, and played the best radiative role. It serves as a major driving force to promote high-quality development of global information technology.

Chinese Academy of Cyberspace Studies, *World Internet Development Report 2019*, https://doi.org/10.1007/978-981-33-6938-2_2

All countries strengthened the plan and guidance of network information technologies (especially cutting-edge technologies), promoted the accelerating agglomeration of high-end innovative resources, deepened the integration of the new-generation network information technologies with other technological fields, and promoted the digital, networking and intelligent transformation of economy and society in deeper and wider fields.

2.2 Basic Network Information Technologies

Basic technologies lie in the bottom of the network information technology ecosystem and serve as the cornerstone. Now, 5G, Internet of Things and artificial intelligence are applied in more fields. It promoted basic technologies such as high performance computing (HPC), software technology and integrated circuit to make significant progress through innovating structural system, promoting integration across disciplines and industries, and optimizing procedures, and to play a comprehensive role in supporting the basic operation and sustainable development of Internet.

2.2.1 High Performance Computing (HPC) Achieved Sustainable Innovative Development

2.2.1.1 The International Competition Pattern of High Performance Computing Remains Relatively Stable

High performance computing is the core of the new-generation information technology and plays a pivotal role in promoting the comprehensive national strength. As the strategic role of information technology has been highlighted, international competition on high performance computing focusing on the R&D of super computer became more and more fierce. In June 2019, "TOP 500", an international organization on super computer performance evaluation, released the latest list of global top 500 super computers, among which the floating point arithmetic capability of the 500th reached 1021 trillion times per second. This signaled that the basic speed of "TOP 500" arrived at thousand trillion per second and that the competition and development of super computer entered a new stage. In the latest list, 219 China's super computers were included, taking the lead over all other countries and regions. The numbers in the United States, Japan, France, the United Kingdom and Germany are 116, 29, 19, 18, and 14 respectively. Comparing with the last list, the number distribution of super computer remained stable. The United States, China, European Union and Japan maintained their advantages.

According to the Performance in "TOP 500 Full List", the United States took the lead in high performance computer, with the largest (half of top ten countries) overall operational capacity percentage of 38.5%. Summit Super Computer, jointly developed by IBM and Nvidia, remained the first (for three times), with a floating point arithmetic speed of 148,600 trillion times per second and the summit of 200,800 trillion times per second. China ranked the second in number of high-performance computing systems, with an overall operational capacity percentage of 29.9%. In Europe, one high performance computing system has been added than that in 2018. Among them, Switzerland's Piz Daint and Germany's SuperMUC-NG ranked the sixth and eighth respectively. In Japan, only ABCI (AI Bridging Cloud Infrastructure) at National Institute of Advanced Industrial Science and Technology (AIST) ranked among the top ten, but it dropped to the eighth (from seventh) in the rank of system performance. See Table 2.1 for global super computer top ten in June 2019.

2.2.1.2 Container Technologies Has Been More Widely Used in Supporting Advanced Computing

With the further development of container technologies represented by Docker and Kubernetes, the ability of container cloud to encapsulate, manage and schedule all kinds of hardware resources in cluster has been greatly improved, which effectively promoted the safety, efficiency and flexibility of advanced computing and established a good foundation for scale deployment. For instance, Tencent has launched TKE (Tencent Kubernetes Engine), a container cloud platform for enterprises. With functions of centralized cluster management and security, TKE can support the deployment of virtual machines, bare servers and graphics processor (GPU) servers in a single cluster, and connect to the container service on the cloud to get consistent management experience. In July 2019, Tencent cloud container product was appraised as "powerful competitor" by Forrester, an international authoritative consulting agency, and listed in global container manufacturers.

2.2.1.3 Advanced Computing Technologies Accelerate the Development Towards Intelligent Computing

As one of the main application scenarios of advanced computing, intelligent computing can accelerate traditional scientific calculations. For example, in April 2019, "Event Horizon Telescope Collaboration (EHT) project team" took the first photo of black hole in human history. The research team used new machine learning algorithm to help create and improve new models and increase calculation speed, which promoted the research on black hole.[1] In addition, the United States super computer Summit has achieved the double-precision floating point arithmetic speed of 148,600 trillion times per second. Yet, for AI applications, half-precision floating point arithmetic can meet the requirements in many application

[1]https://edition.cnn.com/2019/09/05/world/black-hole-photo-prize-scn-trnd/index.html.

Table 2.1 Global super computer top ten in June 2019

Rank	Organization	Country	System	CPU	Floating point arithmetic speed (Time/S)	Summit arithmetic speed (Time/S)	Power consumption (/kW)
1	Oak Ridge National Laboratory	The United States	Summit	2,414,592	148,600.00	200,794.90	10,096
2	Lawrence Livermore National Laboratory	The United States	Sierra	1,572,480	94,640.00	125,712.00	7438
3	National Supercomputing Center in Wuxi	China	Sunway Taihu Light	10,649,600	93,014.59	125,435.90	15,371
4	National Supercomputing Center in Guangzhou	China	Tianhe-2 A	4,981,760	61,444.50	100,678.70	18,482
Rank	Organization	Country	System	CPU	Floating point arithmetic speed	Summit	Power consumption (/kW)
5	Texas Advanced Computing Center	The United States	Frontera	448,448	23,516.40	38,745.91	–
6	Swiss National Supercomputing Centre	Switzerland	Piz Daint	387,872	21,230.00	27,154.30	2384

(continued)

Table 2.1 (continued)

Rank	Organization	Country	System	CPU	Floating point arithmetic speed (Time/S)	Summit arithmetic speed (Time/S)	Power consumption (/kW)
7	Los Alamos National Laboratory, Sandia National Laboratories	The United States	Trinity	979,072	20,158.70	41,461.15	7578
8	National Institute of Advanced Industrial Science and Technology	Japan	ABCI	391,680	19,880.00	32,576.63	1649
9	Leibniz Supercomputing Centre	Germany	SuperMUC-NG	305,856	19,476.60	26,873.86	–
10	Los Alamos National Laboratory, Sandia National Laboratories	The United States	Lassen	288,288	18,200.00	23,047.20	–

scenarios and improve arithmetic efficiency with enormously reduced data transmission and storage. Therefore, a team from Oak Ridge National Laboratory used a testing program called "HPL-AI" to achieve 445,000 trillion times calculation per second, which surpassed HPL performance testing program that was considered a model for high performance computer.

2.2.1.4 Higher Performance Neuromorphic Chip Has Been Released

For a long time, researchers used neuromorphic chip to simulate the function of human brain and promote the development of brain-like computing. In September 2019, Zhejiang University released "Darwin 2", a spiking neuron network brain-like chip, as well as the tool chain and micro operating system for the chip. "Darwin 2" adopted 55 nm technique with more than 10 million nerve synapses, which displayed unique advantages in smart IoT.[2] In August 2019, Tsinghua University released "Tianjic", the first hybrid brain-like computing chip in the world. Based on computer science and neuroscience, it formed a platform and effectively promoted the research and application of artificial general intelligence.[3] Intel launched Pohoiki Beach, a brain-like chip system. It can increase the speed of AI algorithm by 1000 times and the efficiency by 10,000 times. And it can be used in scenarios of autopilot, robot skin and prosthetic.

2.2.2 Software Technologies Are Accelerating the Development Towards Platform

At present, software technologies are accelerating the extension from single machine to network, from terminal to platform. Operating system and industrial software play more basic and strategic role. Related innovative researches and development are continuously deepening. As SDX became an important development direction, related researches are constantly deployed.

2.2.2.1 Operating Systems Are Diversified

As the devices of cloud platform and IoT became diverse, the basic role of operating system was further deepened and consolidated. It is increasingly urgent to develop new operating system according to application scenarios, with new R&D achievements made on terminal operating system, cloud operating system, and IoT operating system.

---

[2] https://news.cctv.com/2019/08/26/ARTIhlnbRuwHuHVUx0kp70QR190826.shtml.

[3] https://tech.huanqiu.com/gallery/9CaKrnQhVHr.

(1) In terminal operating system, new progresses were made in both mobile and desktop ends. In mobile end operating system, Android and IOS rank the first and the second, with other system accounting for less than 2.5%.[4] Some new operating systems also drew attentions. For example, Fuchsia, Google's new operating system, can work without the operating environment of Linux and Java, and be used in a variety of intelligent terminals as a cross-platform system. Russia releases "Aurora", an operating system based on Sailfish, an open source operating system. It concentrates on information security and applies in Russian government and state-owned departments. India developed KaiOS, a mobile end operating system based on Linux. It raises less requirements for hardware and has potential in less developed areas. In desktop end operating system, Windows and Apple have market share of 77.61% and 13.17% respectively, which remain stable.[5] In general, Linux has optimized its released editions in daily use and speeded up the development of ecology aiming at ordinary users according to various application fields.
(2) Cloud operating system is developing towards large scale and cluster. Priority of various cloud applications and services is such factors as reliability, stability and security. Nowadays, major cloud service providers developed cross-platform cloud operating systems based on Kubernetes, an open source container coding engine that belongs to Cloud Native Computing Foundation (CNCF). This greatly accelerated the deployment of software and reduced the difficulty of software deployment in various circumstances, which can probably change the operation and maintenance logic of software industry in a fundamental way. On the development of cloud platform, Serverless architecture has become a new hot topic, with several excellent frameworks, such as Serverless Framework, ZEIT Now and Apex formed.
(3) The R&D and deployment of IoT operating system is accelerating. Emerging operating systems, such as Internet of Things and Internet of Vehicles, attracted many enterprises and institutions to follow up and thrived. More and more computing tasks are deployed in terminal devices, and IoT operating system plays a determinative role to determine the economy. Currently, major IoT operating systems include LiteOS, AliOS Things, Embedded Linux, FreeRTOS and Mbed OS. Meanwhile, as IoT operating systems requires less in technology and stresses "light weight", problems of security and fragmented economy are increasingly prominent.

2.2.2.2 The Idea "Software Defined Anything (SDX)" Is Still Growing

In recent years, as Software Defined Storage (SDS), Software Defined Network (SDN), Software Defined Compute (SDC), and Software Defined Data Center (SDDC) integrated and developed, the idea "SDX" got mature. Its core lies in that software plays a major role in controlling hardware and effectively improves the

[4]https://gs.statcounter.com/os-market-share/desktop/worldwide.

[5]https://gs.statcounter.com/os-market-share/desktop/worldwide.

utilization of resources through the flexible use of functions. Currently, "SDX" has been remarkably used and developed in some fields. In particular, the hardware virtualization technology of large-scale cloud platform enables software staff not to interfere with any hardware performance and accelerates the development and deployment of software applications.

(1) In terms of storage, Huawei FusionStorage is the first data-center level fusion storage in this circle. It supports large-scale horizontal extension, and the system can be easily extended to 1000 nodes and EB-level capacity. IBM Spectrum Storage (product set) can identify and compile unstructured data, expand more than 400 different storage systems horizontally, and quickly connect to cloud. AMD's StoreMI storage system can unify solid state disk and mechanical hard disk into a single virtual hard disk in personal computer.
(2) In terms of transport, China Whale Cloud Co., Ltd. released the first software defined traffic signal system in July 2019, which consists of Road Side Unit (RSU) and Signal Center (SC), a central information control software platform. It can separate the control and execution with the concept of SDX, and provide infrastructure support for the implementation of driverless cars in the future.
(3) In terms of aerospace, the more freely idea "Software Defined Satellite" came into existence. Unlike traditional satellites designed to perform specified tasks prior to launching, "Software Defined Satellite" can perform different tasks according to different needs. In late 2008, China launched Tianzhi I, the first "Software Defined Satellite" in the world. In March 2019, Lockheed Martin Space Systems Company, an American aerospace manufacturer, also launched "Software Defined Satellite".[6]

2.2.2.3 Industrial Software Displays Strength in Competition

As an important foundation and core support of intelligent manufacturing, industrial software plays a strategic role in promoting the transformation and upgrading of manufacturing industry and achieving high-quality development. Among them, important industrial softwares include electronic design automation and computer-aided software.

(1) Several corporations dominate the field of electronic design automation (EDA). Cadence, Mentor and Synopsys divide most of the market share. In June 2019, Cadence launched Tensilica Vision Q7, a new digital signal processor, which can perform up to 1.82 trillion times per second. Used in high-demand visual and artificial intelligence processing applications, it helps to actualize the projects in edge computing. In March 2019, Mentor proposed the idea of EDA 4.0, which aimed to promote artificial intelligence and technological innovation with the help of complete tools integrating circuit to system. In July 2019, Synopsys released a passed yield learning platform for more advanced processes, which laid foundation for the follow-up 5, 4 and 3 nm level massive

[6]Data Source: Website of Lockheed Martin, https://news.lockheedmartin.com/.

production passed yield. Generally speaking, three EDA software corporations in the world will remain the monopolistic role for a long time, and their expensive copyright license transfer fees became a bottleneck restricting the early development of newly-founded chip enterprises.

(2) In the field of open source EDA, evolutions are about to come out. Presently, Chisel tool released by UC Berkeley based on RISC-V open source instruction set is developing rapidly. It's expected to break the monopoly of business EDA. EasyEDA, China's open source free software, is drawing wide attentions soon, and it provides both desktop version and cooperative office cloud version. Other open source EDA software include KiCad, gEDA, Magic and QElectroTech.

(3) The "light weight" of computer-aided software and the integration of industrial design processes is an important trend and considerable achievements were made in many fields. Autodesk used Fusion 360 to integrate all products and created a full-process-industrial design assembly line. For example, Revit AutoCAD can draw both two-dimensional and three-dimensional models. Maya and Arnold can realize rendering animation and visual effects. UGNX, a professional industrial design software of Siemens, enhanced the function integrating design, simulation and manufacturing, and it supports remote collaborative work. ANSYS, an old American simulation design company, worked with French AVSimulation, and released immersion autopilot simulation software, in order to promote the rapid implementation of driverless technology. Previously, French Dassault's CATIAxDesign software released an online light weight version, which marked that traditional professional design software was deployed to the cloud. The AR and VR design functions were introduced in Dassault's SolidWorks 2019. Adobe has launched a range of products driven entirely by AI technology. For instance, Adobe Sensei can dynamically remove video images and convert static images into small videos. Project Kazoo can convert the melody of humming into sound of musical instrument. As most popular game engines in the game development market, Unity and Unreal Engine 4 have released engine tools that support virtual reality and hybrid reality respectively, to realize the rapid development of AR/VR games.

2.2.3 Integrated Circuit Technologies Are Developing Steadily

As the core of network information technology, integrated circuit manufacturing plays a strategic, basic and leading role. It involves such processes as design, manufacturing, sealing, equipment, components and parts, and materials. In 2019, remarkable progress was made in basic chips, advanced technologies, and open source hardware, which inspired new vitality in the development of network information technologies.

2.2.3.1 New Architectures and Technologies of Basic Chips Keep Emerging

Chip is hailed as the "heart of computer" and the "soul of modern information technologies". As integrated circuit gradually approached the physical extreme, chip density and process development have been slower than the earlier prediction of Moore's Law. Yet, innovation on new architectures and technologies enables chip's performance to improve rapidly.

1. Large corporations lead the R&D of computing chips and monopolize the market.
   Computing chips mainly include CPU, GPU and DSP.
   (1) In CPU chip, global market scale achieved $54.15 billion in 2018, with a year-on-year increase of 12.6%.[7] In the general CPU market, X86 processors of Intel and AMD accounted for 96% of the market. Intel's latest i9 series processors are for 14 nm level process. In addition, AMD is catching up with Intel in terms of performance and power consumption through its 7 nm process ZEN2 architecture processor.
   (2) In GPU chip, as the use of more and more large-scale parallel algorithm for deep learning algorithm developed quickly, Nvidia, AMD and Intel maintained their dominant role in global market. With the emergence of special AI chips such as Google's TPU, GPU's share in AI field may continue to decline in the future.
   (3) In DSP chip, global market scale achieved $1.45 billion in 2018, with a year-on-year decrease of 0.1%. Texas Instruments (41.7%), NXP (22.6%), and ADI (21.6%) dominated the global market.[8]
2. The development of memory chips slowed down and the use of new achievements speeded up.
   DRAM and NAND Flash are the main components of memory chip industry, which account for 97% of the market. In 2018, DRAM and NAND declined in both scale and product price. Top 5 memory chip corporations, namely, Samsung, Hynix, Micron, Toshiba, and Western Digital monopolized the market and accounted for 92%.[9] In DRAM, Samsung started to produce 10 nm process LPDDR5 chips based on EUV in 2019. 1 Ynm massive production in NAND process got mature and the next-generation 1Znm is tested now. It is predicted that 3D NAND technology will become a new direction of development after the process enters 14 nm.
3. Communication chips meet new opportunities in 5G age.
   As 5G age arrived, the development of mobile Internet and IoT raised higher requirements on network coverage, connection number, transmission speed and

---

[7]Statistical report of Gartner: *Marker Share: Semiconductors* by End Market, Worldwide, 2018.
[8]Statistical report of Gartner: *Marker Share: Semiconductors* by End Market, Worldwide, 2018.
[9]Statistical report of Gartner: *Marker Share: Semiconductors* by End Market, Worldwide, 2018.

transmission delay, and the performance of communication chips is put on the agenda. At present, major baseband chips are based on multi-mode multi-frequency baseband chip technology. It is an important development trend to integrate baseband chip and computing chip into a SoC chip. Now, there are four 5G baseband chip producers, namely, Qualcomm X50, Huawei Balong 5000, MTK M70 and UNISOC Chuteng 510. RF chips show the following trends.

(1) Compound semiconductors will be widely used in RF devices. In 5G age, base stations will mainly adopt GaN power amplifier that can process UHF millimeter wave above 50 GHz and support high bandwidth; yet mobile phones will mainly adopt GaAs power amplifier.
(2) BAW filter will gradually replace SAW filter depending on the performance advantage of 30–60 frequency band.
(3) RF chips will develop towards integration. In order to meet the requirements of small-scale, thin-light development, RF chips will integrate PA, LNA, switch and duplexer in the future.

2.2.3.2 Advanced Processes Are Approaching the Physical Extreme

New processes and technologies are improved continuously, which enables hardware performance to keep a high-speed growth. Now, the world's most advanced massive production process has been promoted to 7 nm, and great breakthrough has been made in 5 nm level process industrialization. It is expected to reach 3 nm level process. With the reduction of process, the investment into advanced manufacturing production line has increased considerably, and some advanced process manufacturing enterprises have withdrawn from competition. As Global Foundries, the world's second largest Original Equipment Manufacture enterprise, announced the suspension of 7 nm FinFET advanced process research, and UMC, the world's third largest Original Equipment Manufacture enterprise, announced the abandonment of less-than-12 nm process development, competition in advanced process will be mainly among tsmc, Samsung and Intel in the future.

Mask Aligner is the key equipment for chip manufacturing and determines the feature size of the entire integrated circuit process. At present, the 7 nm chip in massive production mainly adopts EUV technology. In recent years, etching equipment has developed steadily, and atomic layer etching technology that can control the accuracy at 0.4 nm has been developed. Lam Research, Tokyo Electron, and Applied Materials account for about 90% of the global etching equipment market. Membrane deposition equipments account for about 22% of the total manufacturing equipments. Now, the United States, Europe and Japan take the lead. The main manufacturers include Applied Materials (the United States), Lam Semiconductor (the United States), Tokyo Electron (Japan), and ASML (Netherlands).[10]

[10]Statistical report of Gartner: *Marker Share: Semiconductors* by End Market, Worldwide, 2018.

2.2.3.3 Open Source Hardware Inspired New Vitality in the Development of Chips

Open source hardware can lower the barriers of hardware development and provide more access to participants. Currently, open source hardwares with wide attention are mainly in the fields of RISC-V and MIPS. Driven by RISC-V Open Source Foundation, RISC-V Instruction Set is developing fast. In controlling field and IoT case, there are more and more RISC-V-based products and application cases. A growing number of open source communities and enterprises are engaged in RISC-V-related adjustment and optimization, and trained many technical personnel familiar with RISC-V architecture in RISC-V frontier research. MIPS Instruction Set is one of the earliest commercial RISC. It once enjoyed equal popularity with X86 and ARM. With the rise of mobile Internet, MIPS Instruction Set gradually declined. On December 17, 2018, Wave Computing, the acquiring company of MIPS, announced MIPS Release 6 (R6), the latest open source instruction set. Besides, IBM opened the source of its Power architecture in September 2019, which further expanded the application of open source hardware in the field of high-performance computing. As open source hardware got diversified and improved, the previous monopoly of international giants has been gradually broken.

2.3 Cutting-Edge Technologies

To a large extent, cutting-edge technologies represent the preemptive advantage and dominant position in the field of science and technology, and technological innovation in the field of network information plays a key role in seizing competitive advantage. All countries enhanced the plan and support in cutting-edge and transformational technological innovation, invested lots of resources, constantly made new breakthroughs, and promoted the iteration and evolution of overall technological ecology.

2.3.1 AI Technology Is Thriving

Artificial Intelligence is a leading and strategic technology in the round of scientific and technological revolution and industrial transformation. It displays distinctive "Leader Goose" Effect. Now, new-generation AI is thriving in the world and driving the world from Internet information age to intelligent information age. As new algorithms and data structures were realized, AI cutting-edge technologies are transforming rapidly. Corporations in all countries make continuous efforts to develop the basic frameworks of AI technology such as chip, algorithm and software, and to be the leader in the field of AI.

2.3.1.1 AI Chip Develops Towards General Chip

Leading information corporations intensively release general AI chips and computing platforms to actualize the implementation of the accelerated computing and application of AI algorithm. In November 2018, Amazon released Inferentia, a new processor chip specially designed for machine learning and reasoning, which supported the frameworks of deep learning and the model of ONNX formats, such as TensorFlow, Apache MXNet and PyTorch. Inferentia can provide high throughput and low-latency reasoning at very low cost and enable complex models to predict quickly. In January 2019, Intel released NNP-1, the latest model of Nervana Series neural network processors, which applies to the acceleration of enterprise-level high-load reasoning task. In March 2019, Facebook released and opened source three AI hardwares, namely, Zion, a Next-Generation hardware platform for AI model training, Kings Canyon, specially designed integrated circuit for the optimization of AI reasoning, and Mount Shasta, for video transcoding. In April 2019, Qualcomm released Cloud AI 100, an accelerator specially for AI. It can enable trained systems to predict new data and can be used to identify new data objects not involved in the training process.

Meanwhile, in terms of special-use AI chips, Tesla released Autopilot 3.0, a self-developed chip for specific scenario autonomous driving, which optimized the performance for massive image and video processing tasks. It can process the image input of 2100 frames per second produced by 8 cameras simultaneously, or 2.5 billion pixels per second, with an increase in performance by 21 times of the last version.

2.3.1.2 Breakthroughs Have Been Made in AI Algorithm

In efficiency and accuracy, AI algorithm kept optimizing and improving the existing algorithm, and made achievements in new fields and algorithms constantly.

(1) Major breakthroughs have been made in the field of Natural Language Processing (NLP). BERT model released by Google is hailed as a milestone event. Different from the previous language representation model, BERT aims to pre-train deep bidirectional representation based on the left and right context at all layers. It can adjust to a small extent with one additional output layer and create the current optimal model for multiple tasks such as question and answer and language reasoning, without great modifications to the specific architecture of tasks. The rise of BERT refreshed the optimal index of multiple NLP tasks and provided new direction for the development of NLP.

(2) New achievements have been made in 3D shape completion. MIT proposed ShapeHD, which surpassed the limits of the existing single-view shape completion and re-shaping techniques by combining deep generative model with the shape prior feature of adversarial learning. Experiments proved that ShapeHD displays the highest level in shape completion and re-shaping of multiple real data sets. The progress of 3D shape completion will promote development in many AI-related fields such as VR, AR and robot.

(3) 6D pose estimation plays a key role in major important practical applications, such as robot grab and control, automatic navigation and augmented reality. A research team from Stanford University proposed an end-to-end deep learning approach which can enable the models to figure out partial shape and geometric information by imbedding and integrating RGB values and point cloud in each pixel. It can conduct 6D pose estimation on known objects input by RGB-D and help to solve heavy shading.

2.3.1.3 AI Software is Developing Towards Deep Neural Learning Framework

Google and Amazon have opened the source of several machine learning platforms and frameworks. It includes the upgraded original frameworks, the frameworks developed for specific platforms, and friendly frameworks for emerging Graph Neural Networks (GNN).

(1) Deep learning is more widely researched and used in mobile end. Owing to the popularity of mobile devices such as smart phones, the application of AI technology in mobile end made more users realize the help of AI in traditional work and daily life, which became the important direction for the landing of AI products. Subject to computing power, deep learning cannot support the network structure and the computation built by major frameworks in mobile end in a perfect way. Now, a series of solutions have been put into effect. In October 2018, Facebook opened the source of QNNPACK, a high-performance kernel library. By speeding up the efficiency of such computing as in-depth type convolution, QNNPACK promoted the use of neural network framework. Now it has been integrated into Facebook applications and deployed in billions of devices. Deep learning has made great progress in computer vision and computer graphics. In March 2019, Google released TensorFlow 2.0 Alpha version. It took Eager execution as default priority mode, and any operation can run immediately after the call. In May 2019, Google launched TensorFlow Graphics. Combining computer graphics system and computer vision system, it can use lots of unannotated data to solve the problem of data annotation in complex 3D visual tasks and to help self-supervision training.

(2) There are a few frameworks that directly deal with NLP or sequence modeling. In February 2019, Google opened the source of Lingvo, a sequence modeling framework. As a general deep learning framework developed by TensorFlow, Lingvo concentrates on the sequence modeling approach related to natural language processing, including machine translation, speech recognition and speech synthesis. By sharing common layers between different tasks, it improved the reuse of codes. As NLP technology got mature and AI research got detailed gradually, more needs may arise in various frameworks such as NLP and CV.

(3) Graph neural network became a hot topic, and research on it enlarged the application field of neural network. In subjects such as social networks, knowledge

graph and life engineering, graph neural network can be used to model node relationships, and the frameworks that can realize graph neural network models are demanded. To design "fast and effective" deep neural network, New York University (NYU) and AWS jointly developed Deep Graph Library (DGL) for graph neural network and graph machine learning. DGL can link up with the current major deep learning frameworks, such as PyTorch, MXNet and TensorFlow, and freely transform traditional tensor computing to graphic computing. The emergence of the framework enormously facilitated the researches on graph neural network, as researchers can attach more attention to the researches on model and algorithm with no need to learn new development framework.

2.3.1.4 Real-Use Products on Basic Application Technology Became Diversified

(1) Pattern recognition was put into use in many fields, which facilitated their service in various fields. In March 2019, Google released a mobile end full-neural speech recognizer. Running at character level, it can achieve real-time speech and text input and work offline. In the field of smart healthcare, Google, Microsoft and Amazon took the lead in deployment and made achievements in succession. For instance, Healthcare Bot, Microsoft's chat robot project, adopted natural language processing technology and can deal with subject changing, human-made errors and complex medical problems. It can support the developer's design and expansion function and provide medical institutions with virtual medical assistants and consulting services.

(2) New technological ideas emerged in the field of automatic engineering, and neural network generation is expected to achieve automation. The complexity of deep neural network restricted its use in devices with limited computing resources such as mobile end. To solve this problem, University of Waterloo AI Institute and Darwin AI Corporation put forward a new idea of Generative Synthesis, which automatically generates deep neural network (FermiNets) with efficient network framework by generating machine. Through the experiment description of image classification, semantic segmentation and target detection task, Generative Synthesis can achieve the goal of higher efficiency of modelling, lower computing cost of computing and higher energy efficiency. In the future, Generative Synthesis is expected to develop into a general method and accelerate the formation of deep neural networks in device end edge scenarios.

(3) Storage performance of knowledge engineering was greatly improved and the formation of bigger knowledge graph became possible. In May 2019, eBay opened the source distributed knowledge graph storage Beam. As a type of knowledge graph storage, Beam adopts distributed storage and supports SPARQL-like inquiry. It supports large-scale graph that cannot be effectively stored by a single server, and generally supports more than 10,000 times of data changes per second.

2.3.2 *The Implementation of Edge Computing is Accelerated*

Driven by AI and 5G technologies, edge computing rose rapidly in 2019 and attracted wide attentions of technological summits, think tanks, cutting-edge technology enterprises and experts. According to the data of International Data Corporation (IDC), by 2020, there will be more than 50 billion terminals connected with devices, and more than 50% of the data will need to be analyzed, processed and stored on the edge of network in the future.

2.3.2.1 The Industrial Ecosystem of Edge Computing Has Improved Gradually

In April 2017, Linux Foundation established EdgeX Foundry community, which aimed to create an interoperable, plug-and-play and modular open source IoT edge computing ecosystem for developers to quickly reconstruct and deploy according to their service requirements. In November 2018, Wind River, a leading IoT software provider on critical infrastructure, and CENGN jointly founded a public software warehouse, to provide StarlingX host resources, help developers build cloud infrastructure, and optimize high-performance and low-latency applications for industrial IoT and telecommunications. In December 2018, Edge Computing Consortium (ECC) released *Edge Computing White Book 3.0*, putting forward new edge frame 3.0. It included four development frameworks, namely, real-time computing system, light computing system, intelligent gateway system and intelligent distributed system, covering the service and development from terminal nodes to cloud computing center links. In January 2019, Edge Computing Consortium Europe (ECCE) was founded to provide a comprehensive edge computing industrial cooperation platform for manufacturers and organizations of intelligent manufacturing and IoT. Its working goals are to build reference architecture model of edge computing, realize full stack technology of edge computing, identify industrial shortage and evaluate the best practice.

2.3.2.2 Many Leading Corporations Have Launched Edge Computing Platforms

Some tech giants, including Amazon, Microsoft and Google, are actively exploring edge computing. As early as 2017, Amazon collaborated with AWS Greengrass on edge computing. Microsoft released Azure IoT Edge solutions, which extended cloud analytics to edge devices and supported offline use. Google released two new products, Edge TPU (hardware chip) and Cloud IoT Edge (software stack), to help enterprises improve the development of edge networking devices. As more networking device emerged, Nvidia, HP, Huawei, Alibaba, Baidu, UNISOC, Fujitsu, Nokia, Intel, IBM and Cisco have carried out technological deployment. For instance, in

May 2019, Nvidia launched EGX accelerated edge computing platform to help enterprises achieve low-latency artificial intelligence at the edge. It can achieve real-time perception, understanding and execution based on continuous data flows among 5G base stations, warehouses, retail stores, firms and other locations. Alibaba Cloud released Link Edg, its product of edge computing, and realized the integrated development of cloud and edge.[11] Baidu Cloud released its intelligent edge computing product (BIE) and its open source version (OpenEdge), which developed towards edge-cloud integration, space–time insight and data intelligence.[12] In July 2019, KubeEdge, intelligent edge project of Huawei Cloud open source, won Peak Open Source Technological Innovation Award. It was also CNCF's first formal project in the field of intelligent edge.[13]

2.3.2.3 Security and Privacy Design Became One of the Leading Researches of Edge Computing

As a new computing model in the age of Internet of Everything (IoE), edge computing is distributed, as "the first entry of data", with limited computing and storage resources. This results in new security threat in addition to common cyberattacks against information systems. For example, the communication packet in an intelligent is tampered with manufacturing firm, which will delay the control valve and cause equipment damage. For another example, unmanned aerial vehicle is misled by analogue signal and lands outside the control area. In 2019, a series of software and hardware on the security of edge computing were released. For example, Intel joined Confidential Computing Consortium, and opened the source of Intel SGX as, which further accelerated the development of confidential computing. Meanwhile, AMD (the United States) released the second-generation EPYC processor, which adopted 7 nm process. Based on new Zen 2 architecture, it provided reliable security performance for hardware. ARM (UK) and its partner on independent security testing laboratory launched PSA Certified to provide framework of standardized security Internet of Things device design for the industry.

2.3.3 *Big Data Technology Keeps Deepening and Expanding*

In recent years, global data scale has increased to the order of quadrillion or even thousand quadrillion bytes, with a year-on-year increase of more than 50%. The world has entered the age of big data. Governments and international organizations have

[11]2018 Apsara Conference Shenzhen Summit, September 25–27, 2018.

[12]Source: "Three Highlights! Baidu Cloud Intelligence Summit 2019 Focuses on AI Industrialization", *China Daily*, August 19, 2019.

[13]Source: "Direct View of Kunpeng Computing Industry Forum: Huawei Cloud Released the Latest Advance!", *China Daily*, 24 July 2019.

formulated relevant policies to actively promote the development and implementation of big data-related technologies.

2.3.3.1 Cloud Computing Technology Boosts the Construction of Big Data Infrastructure

Big data and cloud computing are inseparable. To excave massive data must rely on powerful cloud computing technology. Directed by Internet corporations such as Google, Amazon and Microsoft, to develop big data applications based on cloud computing architecture platform became a common model. In August 2019, Google partnered with VMware to launch a CloudSimple Solution to provide customers with a mixed cloud environment for all-coverage operation and maintenance security. Meanwhile, Amazon, a cloud-computing giant that dominates the global public cloud market, has launched lots of new machine learning services, including AI services for developers, Amazon SageMaker model and algorithm, automated data labeling and reinforcement learning services, AWS optimized version of TensorFlow, and other commonly used machine learning libraries. Microsoft made an acquisition of jClarity, an open source cloud computing platform to support Microsoft Azure cloud services.

2.3.3.2 Big Data Analysis Technology is Touching Cloud Native Environment

In the past decade, big data analysis has been an important technological trend and one of the most dynamic and innovative fields in IT market. Recently, a number of corporate acquisition and merger show that Hadoop and Spark are losing their role in the field of big data analysis, and that big data analysis ecosystem touching cloud native environment is taking shape. In November 2018, Talend, an American storage service provider, bought Stitch, a self-service cloud data integration service provider, to load data to cloud data warehouse with simpler tools. In 2019, a merger of Cloudera and Hortonworks, two well-known big data software corporations, happened. And MapR, another big data unicorn, was bought by HPE. This ended the age of big data business software. Meanwhile, cloud native big data architecture developed fast. Kubernetes, the open source software, as a container that can easily transfer the data between cloud and local data centers, became the foundation of new-generation of cloud native big data. The implementation of containerization projects in sub-fields such as Spark, TensorFlow, streaming media, distributed object storage and block storage made the entire big data stack more flexible to deploy and manage in Kubernetes-based DevOps environment.

2.3.3.3 Big Data Visualization is Developing Towards Interactive Way

Big data visualization and visual analysis can simplify and refine data flow quickly and effectively, and help users to interactively filter the massive data faster and better. Data visualization tools must be upgraded real-time, easily operated, and presented in a multi-dimensional way. And it must support multiple data sources. Currently, the main tools include follows: Tableau software provides a flexible and dynamic console that can monitor information and give complete analysis. QlikView software is a competitor of Tableau. It enables various terminal users to interactively analyze important business information in a highly visual, powerful and creative manner. Microsoft Power BI can connect hundreds of data sources, simplify data preparation and provide specialized analysis. It can realize extension, built-in management and security functions within enterprises.

At present, the innovation of big data visualization is to embed machine learning into business analysis, and merger and integration in the field occur frequently. For example, Google made an acquisition of Looker, a platform for business intelligence software and big data analysis; Salesforce, an American customer-relationship management software service provider, made an acquisition of Tableau; Alteryx, a self-service data analysis software service provider, made an acquisition of ClearStory Data, a big data processing service provider; etc.

2.3.4 Virtual Reality Faces New Starting Point

After the rudimentary, developmental and reflective stages, virtual reality technology and industrial ecology have been constantly improved, and their subversive role was emerging. “2019, New Starting Point” became a hot word in virtual reality industry.

2.3.4.1 Virtual Reality Displayed New Forms and Features

Virtual reality is featured with heavy immersion, interaction, imagination and intelligence. With the development of sectionalization of network information, virtual reality increasingly integrated with other research fields and displayed new forms and features. The multi-sensory output forms, such as “vision”, “hearing” and “touch”, embody how virtual reality works on users. Now, some virtual reality technology evolves into an upward stage of the immersion and multi-channel interaction. Presently, the available major technological indexes are as follows:1.5–2 K monocular resolution, 100°–200° angle of view field, 100 Mkbps, 20 ms MTP time-delay, 4 K/90 Frame Rendering processing capability, inside to outside tracking and location, and immersive sound.[14]

[14]Chen Xi, The Development of Virtual Reality Presents New Trends, www.cnii.com.cn, People’s Post and Telecommunications, January 24, 2019.

(1) Zoom display became a hot technology and explicit computing emerged. Oculus developed Half Dome, a prototype that adopted zoom display technology, which harmonized vergence-accommodation conflict and solved vertigo problem basically. For the better advanced experience of virtual reality, near-to-eye display can not only present the content, but also perceive the user's state. Based on the new technology of computing inside display, near-to-eye display has great potential for development. Nvidia has developed a prototype for computing inside display.[15]
(2) With the rapid rise of foveated rendering and in-depth learning rendering, it's a trend to develop refined rendering of terminal-cloud coordination and software-hardware coupling. The main contradiction in rendering processing field is between the user's demand for better experience and the deficiency of rendering capability. Better image quality, visual fidelity, rendering efficiency and power consumption are technical drivers of this field. At present, the industry mainly focuses on virtual reality-oriented foveated rendering, in-depth learning rendering and mixed rendering, aiming to explore a way for refined rendering of software-hardware coupling.

2.3.4.2 Technological System on Virtual Reality Has Been Improved

After long-term exploration, theoretical and technological problems on virtual reality have been solved and breakthrough has been made to some extent. Technological theory and system on virtual reality are constantly improved. For example, in acquisition and modeling technology, the research focuses on object digital modeling construction in refinement, facilitation, automation, intelligence and high-efficiency have been confirmed according to the deficiency in industrial application. In multi-source data analysis and utilization technology, semantic computing and efficient reuse of digital content became the current problems. The core of exchange and distribution technology is open content exchange and copyright management technology. The authentication detection technology on digital image and video also became a hot topic. In display and interaction technologies, 3D display device based on computer display, head mounted display, multi-directional virtual image suspension display device, and real 3D display device are the current research directions. Multi-channel interaction is user-centered. Natural, intelligent, efficient human–computer interaction by vision, voice, pose, and expression remains a hard task. Technological standards and evaluation system got mature, covering audio/video source standard, content coding standard and 3D Internet standard. Moving Pictures Experts Group and International Telecommunication Union made great contributions to the standardization of digital audio and video coding and decoding.

At the same time, it is noticeable that virtual reality also faces a series of challenges, such as the input and interaction of head mounted display, the integration of spatial computing and virtual reality and its outdoorization, collection, editing and

[15]Chen Xi, The Development of Virtual Reality Presents New Trends, www.cnii.com.cn, People's Post and Telecommunications, January 24, 2019.

interactive playing of virtual reality video, virtual reality based on mobile terminal and Internet, more performance of physical features and new physical models, and flexibility of force interaction and new natural interaction.

2.3.4.3 Virtual Reality Technology Has a Promising Prospect

As a subversive technology, virtual reality has a good prospect. In the future, it will break the limits of 2D display and realize 3D and real 3D display. It will realize panoramic display and interactive experience, and achieve man–machine natural interaction with hand–eye coordination. It can construct virtual copies of anything in any physical world in spite of space–time limitation and realize a digital parallel world. It will help all industries to launch new test and verification platform. It will replace the existing mode of communication interaction represented by Internet mail system, and become a new entrance of Internet and a new environment for interpersonal interaction. It can be expected that in the future, in the fields of mixed reality (MR), augmented reality (AR) and virtual reality (VR), many new applications, new industrial forms, and new achievements will emerge in the field of virtual reality, through the organic integration of VR technology with mechatronics, robotics, artificial intelligence, computer vision and control engineering.

2.3.5 Innovative Achievements Are Continuously Made in Quantum Information

2.3.5.1 Quantum Communication Technology Has Developed and Got Mature Continuously

The performance of quantum devices has been continuously optimized and the technical specifications have been greatly improved. Taking the quantum storage unit achieved by rare-earth doped optical crystal scheme as an example, its coherence time is as long as 1.3 s,[16] with fidelity up to 99%, and storage efficiency increased to 68%.[17] Quantum communication theory and experiment have advanced steadily in solving complex channel problems. The secured communication distance of decoy state and measurement-device-independent protocols under asymmetric channel has been further improved. TF-QKD theory is proposed, in which long-distance and high key generation rate communication can be achieved without quantum repeater and verified by relevant experiments.

[16]Data source: *Nature Pyhsics*, 2018, vol.14, issue 1.

[17]Data source: *Nature communications*, 2018, vol. 9, issue 1.

2.3.5.2 Development of Quantum Computing Overall Architecture Advanced Remarkably

(1) Superconductivity and Ion trap systems are becoming the ideal hardware physical platforms for large-scale universal quantum computing. D-wave (Canada) pre-displayed a next-generation 5640-superconducting bits annealing simulator based on the modified Pegasus scheme. Its entanglement, correlation and computing capability will be greatly improved. Google (USA) released Bristlecone, a superconducting computing chip with 72 physical bits. USTC team finished the preparation for a 20-superconducting-bit entanglement state. IonQ (the United States) released its latest Ion trap-based quantum computer with 79 processing bits and 160 storage bits. When it runs at 13 bits, the average fidelity of a quantum gate is 98%. It can run at room temperature, and provide important ideas for quantum computing to serve the public out of the laboratory.

(2) Major breakthroughs have been made in the researches related to quantum computing of silicon-based semiconductor system. On the basis of previous researches, the research team from the University of New South Wales (NSW, Australia) raised the fidelity of silicon-based single quantum bit gate to 99.96% and that of double quantum bit gate to 98%, which made a significant progress for the practical researches on error correcting code.

(3) Quantum compilation environment and programming language, quantum cloud services and other software platforms are gradually opened. IBM (the United States) Qiskit, an open source quantum computing framework started to provide cloud access services. Qutech (Netherlands) launched the quantum platform Quantum inspire and provided users with simulation service of 37-bit quantum algorithm. Rigetti (the United States) launched quantum cloud service and announced to release a 128-bit quantum computing system this year. Huawei (China) released HiQ, a quantum simulation platform, and opened the authority of cloud usage to related researchers. Microsoft (the United States) improved the language Q# of quantum programming and released simulation tool library on quantum chemistry. D-wave (Canada) announced to open source of the hybrid workflow platform D-wave Hybrid and provide classical-quantum hybrid programming tools for developers.

(4) Cryptographic security in post-quantum age became an important topic. After a year's test, United States National Institute of Standards and Technology (NIST) selected 26 post-quantum encryption algorithms for the second round of evaluation, including grid-based, coding-based, multivariate structure and other important schemes, to research the problems of safe transmission and processing of information in the post-quantum age.

2.3.5.3 All Countries Have Clarified Their Short-Term Development Priorities in Quantum Computing

Governments of all countries are striving to achieve the maturity and popularity of technologies on OCO simulation, quantum optimization, quantum control in the next few years. National Science Foundation of the United States provided University of Maryland and other six institutions with $15 million for STAQ Project, aiming to research and develop the software/hardware foundation necessary for quantum computer and to explore its application fields and value. National Science Foundation of the United States launched Quantum Computing and Information Science Talent Training Program, with personal discretionary funds up to $750,000. European Union announced OpenSuperQ Program and allocated 10.33 million euros for the related researches. In June 2019, the United Kingdom announced a total investment of more than £1.2 billion to develop quantum computing.

2.3.5.4 New Advance Has Been Made in the Development of Precision Instrument with the Help of Quantum Sensing

Research on quantum sensing technologies for precise measurement of physical quantities by using quantum system, quantum trait and quantum phenomenon has been constantly strengthened, and the measurement accuracy has been continuously improved. The measurement of weak current and weak magnetism became an important application. Universität Stuttgart (Germany) applied Nitrogen Vacancy (NV) color center system in diamond to the accurate measurement of electric field and obtained high measurement accuracy. Other quantum systems, such as optical microcavity system, Rydberg atom and Ion trap system, also displayed their sensing application values in some fields. Quantum sensing-based atomic clock, quantum gyroscope and other devices can be widely used in military fields such as positioning and guidance. And it has become a key field of development in recent years.

As the law of technological development suggests, the world is in the initial stage of the development of network information technologies. Technologies will keep advancing forward, and more subversive changes will take place in the future. It is predictable that major countries, scientific research institutes and leading corporations will make more efforts. In particular, major countries will enhance their plan and layout in the fields of basic, general, cutting-edge and subversive technologies to seize the leading position. In the short term, the fields of technologies such as mobile end operating system, artificial intelligence, Internet of Things, and Internet of Vehicles closely integrated with industries develop faster. The fields such as Big data and virtual reality will get further mature in industry and quantum computing will play more subversive role with the industrial development. According to the development trend in 2019, it can also be predicted that more breakthroughs will emerge in 2020. Government, enterprise and industry need to attach more attention, strengthen international cooperation on scientific and technological innovation, adapt to new technologies in a timely way, apply new technologies effectively, and manage new technologies scientifically, so as to promote high-quality economic and social development.

Chapter 3
Development of World Digital Economy

3.1 Outline

At present, under the growing pressure of global economic downturn, economic development enters a new round of mid-Kondratiev Wave. China displays great economic potential and resilience, the United States boasts the only one among the developed economies expected to gain speed in growth, and India and Vietnam show prosperous outlook among the emerging markets. Yet, across the Eurozone, Japan and other major Asian emerging economies, a slowdown in growth appears. Against this backdrop, countries around the world are seizing the opportunities technological revolution and industrial transformation have produced to vigorously develop their digital economy and to fuel their growth, expand demand, and boost international trade and investments.

The development of digital economy is highly associated with the size of a country's economy, featured by great disparity and underdevelopment, where America stays ahead and strong while developing countries remain less heard. At the G20 Osaka Summit, Bo'ao Forum for Asia, APEC China CEO Forum and other international conventions, digital economy has become an important topic. Countries keep strengthening the strategic layout of their digital economy and gradually improve their strategic design and policy systems. International organizations such as the United Nations Conference on Trade and Development (UNCTAD), OECD, and Internet Society (ISOC) have paid great attention to digital economy and released reports on this matter.

As the cornerstone of digital economy, information and communication technology (ICT) advances steadily around the world, continuously driving the rapid development of digital economy and supporting all aspects of national economy with various information technologies, products and services. As digital dividends dropped, the development of telecommunications industry has slowed down. Internet companies are opening up new businesses to support the fast growth of their revenues, and global public cloud market is rapidly expanding as Internet penetrates further

Chinese Academy of Cyberspace Studies, *World Internet Development Report 2019*, https://doi.org/10.1007/978-981-33-6938-2_3

into industries. The first year of commercial uses of 5G has created related markets on large scales and suggested the flow of investment.

The ever-deepening integration of advanced manufacturing and modern service industries accelerates the release of digital dividends and promotes industrial transformation on a large scale. Artificial intelligence (AI), big data, and Internet of Things (IoT) hold great promise and become "multiplication factors" that magnify productivity. Industrial Internet platform market expands fast, Fintech industry moves into a new stage of healthy development, and artificial intelligence penetrates further into traditional industries such as medical care, manufacturing, and agriculture.

3.2 Development Trend of Global Digital Economy

According to *Digital Economy Report 2019* released by UNCTAD, over the last decade, global export of ICT services and digitally delivered services has grown faster than that of the overall services, indicating the increasing digitalization of the world's economy. In 2018, the export of digitally delivered services reached $2.9 trillion, accounting for 50% of global service export.[1] As a new growth-driver for global economy, digital economy holds an increasingly important position in the economy across all countries.

3.2.1 Bright Prospect

Since international community has not reached an agreement on the definition of digital economy yet, the data collection criteria vary from one another. As a result, it is hard to determine the size, structure and value of digital economy accurately, but it is commonly believed that digital economy holds great prospect. According to the statistics in the *Digital Economy Report 2019* issued by UNCTAD, digital economy was estimated to account for 4.5–15.5% of the world's GDP measured by narrow and broad perspectives. In particular, in terms of the value added in ICT industry, the United States and China account for 40% of the world's total, and in terms of its size as a percentage of GDP, China's Taiwan, Ireland and Malaysia are among the top of the list. *Measuring the Digital Transformation: A Roadmap for the Future* by OECD focuses on the impacts of digital economy on social development, and believes that digital economy plays a positive role in improving access, increasing efficient use, unleashing innovation, securing employment, boosting social prosperity, building trust, and opening up markets. Another OECD report *Skills Outlook 2019: Thriving in a Digital World* reckons that new digital technologies change the way we live, work and study, and create great potential in promoting productivity and benefiting the society. *Global Internet Report: Consolidation in the Internet Economy* issued by

[1]Digital Economy Report 2019 by UNCTAD.

ISOC shows more interested in the future development path of Internet, and points out the "deepening dependence" of platforms in digital economy and the growing consolidation of all key areas of Internet, including platform, interoperability, standard development and delayering infrastructure.

3.2.2 *Divergence in Digital Strength*

In the strength of digital economy, there emerges an increasingly clear three-echelon formation. Digital economy in all countries has grown by varying degrees, resulting in regional disparity, featured by a high correlation between the size of digital economy and the volume of gross economy. As the largest economy in the world, the United States ranks among the top in terms of digital economy size and digitalization of various industries, pulling far ahead in digital strength. China, Japan, Germany, the United Kingdom, France, and South Korea actively develop key areas of their digital economy and closely follow after the United States. India, Brazil, Canada, Italy, Mexico, Russia, Australia, Indonesia and South Africa lie in the third echelon measured by the development of their digital economy.

According to *Digital Economy Report 2019* by UNCTAD, only 20% of the population in the least developed countries has access to Internet, compared with 80% of that in the developed countries. The gap is even wider with digital data use and cutting-edge technology capability, e.g. Africa and Latin America together own less than 5% of the world's co-location and data center services. The gap, if not narrowed, will aggravate the existing problem of income disparity.

3.2.3 *Distinctive Regional Development*

Europe wins out in overall strength of digital economy. Germany, the United Kingdom, France, Italy, and Russia develop neck to neck with slight difference in digital industrialization and industrial digitalization. Asia also scores high in digital industrialization. China, Japan and South Korea value the full-range, all-round, all-chain transformation of traditional industries with new Internet technologies and applications. Digital technology and industrial development have become an important drive for digital economy. North America takes the lead in the average scale of digital economy. Owing to the strength of the United States, North America realizes over $4 trillion in this regard, far ahead of any other regions. Oceania also remains the top in digital industrialization. Australia and other Oceanian countries put great weight on the integrated development of digital economy, in which Australia achieves high level of digital industrialization, up to 83.4%. Africa lags behind with untapped potential in digital economy. Yet, South Africa reaches a growth speed of 19.5% in digital economy, which may help pull neighboring areas forward. Brazil and other Latin American countries fail to present remarkable development in digital economy, just with mediocre performance.

3.3 Global Consensus on the Development of Digital Economy

Currently, global economy faces all kinds of complications and challenges, and trade frictions continue to escalate. Global economic recovery weakens and switches to a lower gear. Digital economy, as a new growth-driver for global economy, draws more attention across all countries that quicken the strategic deployment of digital economy.

3.3.1 Enhancing Planning and Deployment

At present, countries are stepping up their strategic planning of digital economy. The United States focuses on key areas of cutting-edge technology, makes use of high value-added links of the manufacturing chain, and adopts digital technology to facilitate manufacturing revolution and revitalize traditional industries. By formulating a set of comprehensive data laws and rules, European Union pushes the establishment of digital single market and strengthens data resources management to secure the standardized development of digital economy. Based on the strengths of their telecommunications industry, Japan and South Korea set the development of digital industrialization as a priority. China strives to improve its ability to innovate in the field of digital technology, accelerate digital industrialization and industrial digitization, deepen the integration of Internet, big data, and artificial intelligence with real economy, and give full play to the strength of its big market. In general, policies developed and enforced by countries and regions concentrate mainly on enhancing technological innovation, improving digital application, strengthening governance, and developing smart economy. See Fig. 3.1 for digital economy strategies adopted by major countries in the world.

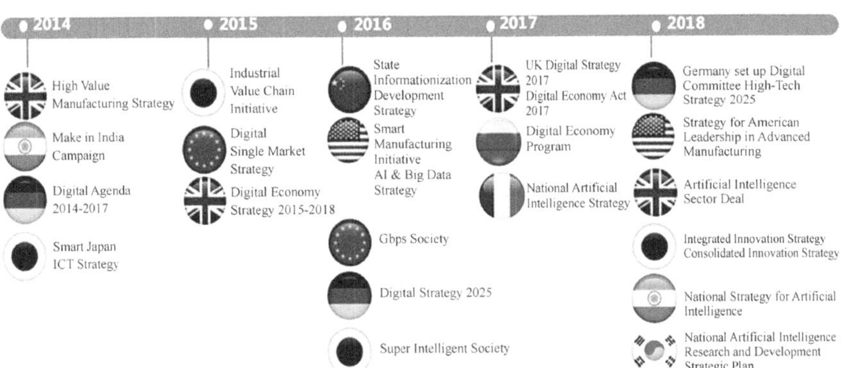

Fig. 3.1 Digital economy strategies of major countries

3.3.1.1 On Enhancing Technological Innovation and Industrial Capability

1. Countries have stepped up their innovation on digital technology, products and services, and actively formulated relevant incentive strategies.

The United States has successively issued *Federal Cloud Computing Strategy, Big Data Research and Development Initiative, Technologies and Policies Supporting Data-Driven Innovation*, which promotes technological innovation from commercial practice to national strategy and secures America's competitiveness in data science and innovation.

Germany has formulated *Digital Strategy 2025*, making the application of digital technology a priority, which is expected to bring in an additional 82 billion euros to German economy by 2020. It builds two big data centers to promote the application of big data innovation in the fields of "Industry 4.0", life sciences and healthcare, and increase the investment in ICT, information security, microelectronics and digital services.

The United Kingdom has issued *Digital Charter* to encourage the growth of local digital technology business and attract technological innovation companies from all over the world to fuel its development.

European Digital Agenda puts forward the concept of "digital technology standard and compatibility" to ensure the seamless interaction between new digital technology devices, applications, data repositories and services.

Japan emphasizes and supports ultra-high-speed network transmission technology, data processing and pattern recognition technology, sensor and robotics technology, software development and nondestructive testing, and multilingual speech translation systems.

Mexico strives to expand its ICT product and service export, aiming to become the world's second largest exporter of IT devices.

2. Broadband network serves as a strategic public infrastructure to support economic and social development.

The United States aims to bump up the actual download speed to at least 100 Mb/s and actual upload speed to at least 50 Mb/s for 100 million households by 2020.

Germany includes its plan in *Digital Agenda* to build a high-speed broadband network across the country with the download speed of over 50 Mb/s by 2018.

The United Kingdom has issued *2017–2019 Telecommunications Infrastructure (Draft)*, which grants tax preference to local businesses with new optical networks for up to 5 years. The budgets submitted by the local government to the Parliament allows up to 60 million pounds of tax relief for businesses that set up new 5G or FTTH/P broadband network, which can be a great support for network expansion.

Canada proposes "Connect Every Canadian" to ensure that residents in rural areas can get access to high-speed broadband networks, fully enjoy cheap wireless services, and participate in and benefit from digital economy.

Norway strives to improve the security and stability of its telecommunication networks by strengthening cooperation among Ministry of Transport Ministry of Communications, suppliers, and post and telecommunications office on cybersecurity.

Japan plans that in the event of natural disaster on a massive scale, the ICT sector can maintain normal function with redundant ICT infrastructure.

3.3.1.2 On Strengthening the Application of Digital Technology

(1) Promoting the integration of digital technology with education, upgrading broadband infrastructure, increasing the number of computer hardware at schools, and enriching online courses. The United States allocates $3.9 billion each year for the construction and modification of broadband networks to ensure that schools and libraries across the country can enjoy high-speed and stable broadband access. The United Kingdom aims to support digital skill learning and labor re-training with massive open online courses (MOOCs).

(2) Promoting the integration of digital technology with transportation and logistics, and using digital technology to create a safe, economic and environmental-friendly road traffic system.

(3) Speeding up the deployment of digital health strategy, applying digital technology into healthcare industry to improve the quality and efficiency of diagnosis and treatment, reduce operating costs, and build new healthcare models. Japan has stepped up the construction of digital infrastructure across medical institutions, and taken various measures, including promoting remote diagnosis and treatment technology, electronic medical records as well as digitalization of prescription and dispensing information, to strengthen knowledge and skills of their medical staff and improve the quality of its medical services.

3.3.1.3 On Improving Governance Capability

(1) Encouraging the construction of digital government. The United States proposes *Open Government Directive*, *Executive Order on Making Open and Machine Readable Government Information*, and *Open Government Partnership: The Second U.S. Open Government National Action Plan*, which expressly demand that all federal government agencies release internal electronic data collection inaccessible to the public at open websites. The United Kingdom describes data as the lifeline of currency innovation and knowledge economy. Japan proposes "Becoming the Most Advanced IT Country in the World" Plan, one aim of which is to allow anyone, at any time and any place, to get access to public sector data and enjoy public services through a one-stop e-government portal.

(2) Guaranteeing the healthy development of IT industry through legislation. The United States attaches great importance to the protection of technology R&D,

patent and intellectual property rights in Internet industry. It establishes a patent system for core and key areas, emphasizing that a sound intellectual property rights protection system facilitates the development of biotechnology, digital technology, Internet and advanced manufacturing industry. In order to strengthen cybersecurity, reduce cyber crimes and protect intellectual property rights of innovative entrepreneurs, British government promulgates *Digital Economy Bill*, stipulating the protection of digital intellectual property rights and digital information in the law. It has also taken active measures to ensure the timeliness, convenience and effectiveness of intellectual property rights protection, thereby encouraging knowledge innovation. China has issued *Instruction on Promoting the Standardized and Healthy Development of Platform Economy* to revise and improve market access requirements for Internet platform economy and encourage the development of new business forms in this field.

3.3.1.4 On Powering Smart Economy

In technological innovation and industrial upgrading, more and more countries concentrate on artificial intelligence. In more than 20 countries where artificial intelligence is basically deployed, more than 16 have intensively released AI strategic plan between 2017 and 2019. Since 2019, the United States and Europe have speeded up their progress, and emerging countries have followed closely.

The United States keeps strengthening strategic guidance and evaluates and adjusts its AI priorities. President Trump signed *Executive Order on Maintaining American Leadership in Artificial Intelligence* in February 2019 and launched "U.S. Artificial Intelligence Initiative". In June 2019, the United States released an update of *National Artificial Intelligence Research and Development Strategic Plan*, changing strategic research and investment priorities from seven to eight.

Members of European Union work collaboratively to advance artificial intelligence and increase their investment in this field. EU Council reviewed and approved *European Coordinated Plan on Artificial Intelligence Development and Use* in February 2019 to promote cooperation among EU members in the four key areas, including increasing investment, data supply, talent training and maintaining trust. In April 2019, European Commission issued *AI Ethics Guidelines* to boost people's trust in artificial intelligence technology products.

Russia, South Korea, Spain and Denmark have stepped up their efforts to formulate national strategies for artificial intelligence. In January 2019, Ministry of Science Technology Information and Communication of South Korea formulated *Plan for Promoting the Development of Data, Artificial Intelligence and Hydrogen Economy*. In March 2019, Spain issued *Spanish Artificial Intelligence Research, Development and Innovation Strategy*, and Denmark issued *Danish National Artificial Intelligence Strategy*.

3.3.2 *Intensifying Fight for the Power to Make International Rules*

Formulating international rules for digital economy serves as an important means to re-define international trade market in information age.

At present, in spite that to make universal rules for digital economy is limited by practical and theoretical conditions internationally, major countries endeavor to develop digital economy by establishing bilateral and regional free trade agreements. Now, these agreements have been accepted by a small circle of countries and are expected to serve as multilateral international rules. For example, in *Japan-EU Economic Partnership Agreement*, it is made a special stipulation for e-commerce and cross-border data flow. In *Comprehensive and Progressive Agreement for Trans-Pacific Partnership (CPTPP)*, an entire chapter is dedicated to e-commerce regulation. In *U.S.-Mexico-Canada Agreement (USMCA)*, detailed digital trade rules are included. According to statistics, there are about 70 *Free Trade Agreements (FTA)* around the world that involve digital trade rules, and there is a growing trend that these bilateral agreements will lead to multilateral agreements.

Although emerging countries have made some progress with digital economy, problems such as disparity in industrial innovation, underdeveloped policies and regulations, and cognitive barrier between traditional industries and new technologies remain unresolved. Digital divide still exists, leaving emerging powers at a disadvantage in getting themselves heard and formulating the rules. Developing countries such as India and Vietnam actively participate in WTO negotiations on e-commerce. Yet, as their standpoints and interests are not in alignment with many other countries, they stand in unfavorable position in making international rules for digital economy.

3.4 Stable Development of Overall Digital Industrialization

As the cornerstone of digital economy, ICT industry provides a wealth of information technology, products and services for various fields of national economy. Currently, ICT industry displays sizable scale and takes up a considerable proportion of GDP, playing a basic role in the development of digital economy.

3.4.1 *Telecommunications Industry Slowing Down Again*

The dividend of 4G traffic dries up now. The three quarters of 2018 witnessed a 0.4% drop in global mobile service market, with over half of the countries/regions experiencing a slowdown in the growth of their mobile service market. Owing to the adverse effect of the decrease in traffic value and weak market, more and more leading

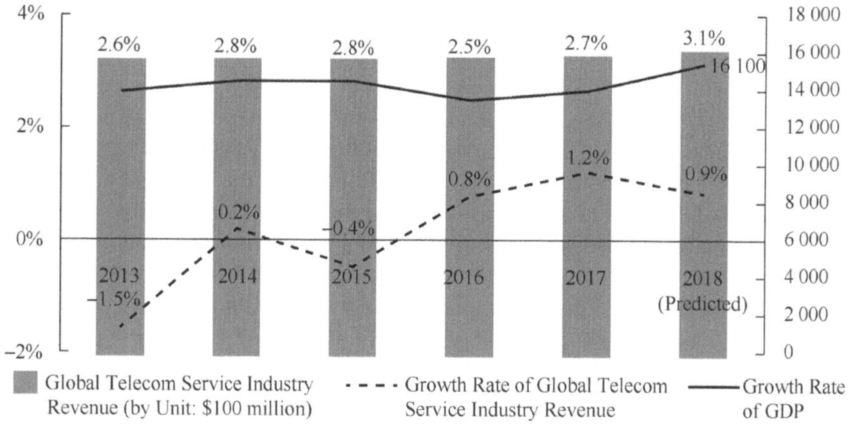

Fig. 3.2 Development of Global Telecom Service Industry from 2013 to 2018. *Data Source* Data center of China Academy of Information and Communications Technology (CAICT), Gartner, IMF

telecom operators fail to keep their revenue in the black. AT&T and CenturyLink rely on acquisitions to drive their growth.

The growth of global telecom industry falls behind the growth of GDP. In 2018, the expected revenue of global telecom service industry was $1610 billion, with the growth rate falling to 0.9%. Specifically, data traffic business served as the main driver of growth, accounting for 60.3% of the revenue in 2018, up from 58% in 2017. However, its contribution rate dropped from 266% in 2017 to 194% in 2018. See Fig. 3.2 for development of global telecom service industry from 2013 to 2018.

3.4.2 *Ever-Growing Electronics and Information Industry*

Global electronics and information industry market grew in 2018.

(1) Smart phone market continues to expand. Technological innovation such as full screen and fingerprint identification results in new-generation smart phones and a rise in price. In addition, the prices of memory chip, capacitor and other components, known as the upstream of the industry, continue to rise, driving up the cost and price of smart phones and subsequently enlarging the smart phone market.

(2) Electronic component market undergoes significant differentiation. Global supply of discrete devices like MLCC (multilayer ceramic capacitor) remains low, which continues to push up the price. New technologies like artificial intelligence drive the growth of optoelectronic devices and FPGA (field programmable gate array). The price of NAND (a flash memory device) has started to fall and the price of DRAM (dynamic random access memory) has

dropped since the second quarter of 2018, leading to the slowdown in the overall growth of electronic component market.

(3) Server and PC market achieves substantial growth. As large-scale data centers enter procurement cycle, the resulting wide-range deployment of IT devices gives server industry a strong push. In addition, PC market starts to recover, forming a new growth-driver. See Fig. 3.3 for main source of revenue growth for global electronics and information industry from 2017 to 2018.

(4) Revenue in software industry continues to grow. Specifically, robotic process automation (RPA) software develops rapidly. According to the data Gartner released, in 2018, corporate application software around the world generated a total of $193.6 billion in revenue, a 12.5% increase year on year; the revenue of RPA software soared by 63.1%, making it the fastest growing part of global corporate software market.

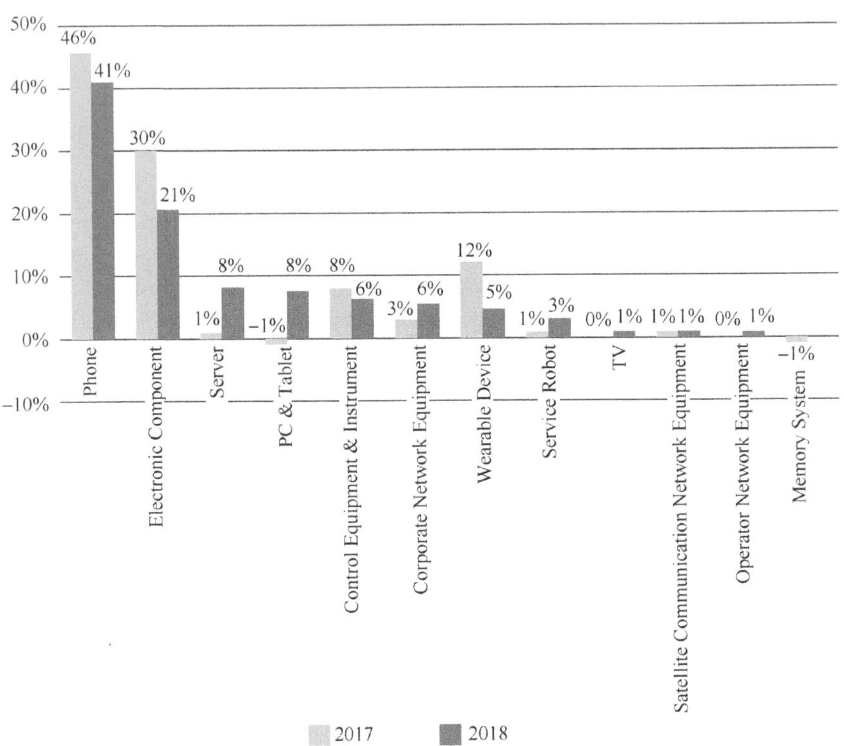

Fig. 3.3 Main source of revenue growth for global electronics and information industry from 2017 to 2018. *Data Source* China Academy of Information and Communications Technology (CAICT), Gartner, yearbook of world electronics data, IFR etc.

3.4.3 Ever-Growing Revenue in Internet Business

On the list of the world's top ten corporations measured by market value, Internet corporations account for nearly 50% the total value. The number of ICT corporations in the list grew from four in 2009 to eight in 2018, with their combined market value rising from 37.9 to 85.8% of the total. Global Internet operating revenue continues to rise rapidly. According to 2018 performance reports of major Internet corporations, the top ten Internet corporations with the highest operating revenue in the world (see Table 3.1) produced a total revenue of $656.1 billion. Among them, Amazon ranked the first with $232.9 billion (approximately ¥1.61 trillion) in operating revenue, Alphabet (the parent company of Google) came second with $136.8 billion (approximately ¥0.94 trillion), and JD came third with $67.2 billion (approximately ¥465 billion). In the top ten list, the United States occupies six and China occupies four, indicating their leading role in Internet field.

3.4.4 Rapid Rise of Public Cloud Market

Global corporate Internet application service market maintains stable growth at 30%. According to Canalys, a well-known market research institution, global public cloud computing market exceeded $160 billion in 2018 and was expected to exceed $190 billion in 2019. In the development of corporate Internet, the United States has been taking the lead in global cloud market and fully engaged in innovation in this field, no matter giants or start-ups. Amazon remains the industry leader, with its market share far ahead of Microsoft, Google and Alibaba. Salesforce, a SaaS unicorn, realizes a market value of over $100 billion. Google Cloud Services also sustains rapid growth, vigorously moves towards new corporate retail market, and enters the field of smart city that serves the government. Chinese Internet corporations strongly compete in Business-end and Government-end markets. Leading corporations such as Baidu, Alibaba and Tencent have been fully engaged in structural adjustment and business layout. Smart city, industrial Internet, artificial intelligence and data analysis become the core business of current Internet corporations. See Fig. 3.4 for scale of global public cloud market.

3.4.5 5G Driving the Development of Upstream and Downstream Industries

As Qualcomm predicts, by 2035, 5G will create $12.3 trillion of economic output around the world, with its value chain alone generating $3.5 trillion in output. From 2020 to 2035, the contribution of 5G to global GDP's growth will be equivalent to the

Table 3.1 World's top ten internet corporations with the highest operating revenue in 2018

Rank	Corporation	2018 (fiscal year) operating revenue/$100 million	2018 (fiscal year) net profit/$100 million
1	Amazon	2329	101
2	Alphabet	1368	307
3	JD	672	5
4	Alibaba	562	139
5	Facebook	558	221
6	Tencent	466	117
7	Netflix	158	12
8	PayPal	154	21
9	Baidu	149	41
10	Booking holdings	145	40

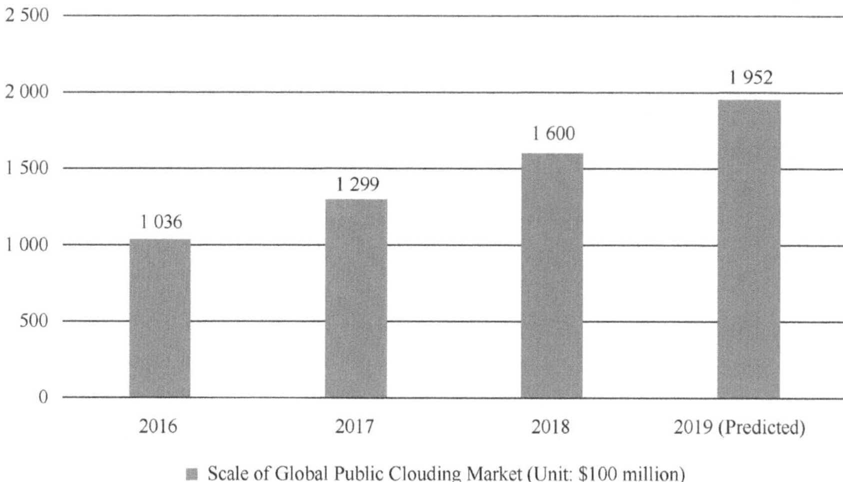

Fig. 3.4 Scale of global public cloud market. *Data Source* Calculation based on IDC data

size of India's economy. According to China Academy of Information and Communications Technology, it is estimated that from 2020 to 2025, 5G will boost a growth in China's digital economy by ¥15.2 trillion. The integration and development of 5G and new ICT technologies such as artificial intelligence and big data will facilitate the profound revolution of digital economy in terms of production organization, resource allocation efficiency and management service model.

5G system and equipment mature and cause the upgrading of communication components and technologies such as radio frequency, antennas and optical modules, as well as more industrial demand. In regard of basic hardware like integrated circuit, the increase of frequency bands and massive connected devices in 5G age give rise to significant growth of filters and power amplifiers. By 2020, global stock of RF devices will reach $20 billion and 200 mm equivalent RF circuit SOI wafers will exceed 2 million pieces in sales. Besides, change in 5G network architecture and deployment in base station will effectively stimulate the demand for optical modules. The investment in China's optical module industry (including wireless network and transmission network) is estimated to reach ¥150–170 billion in 5G age. In regard of equipment, Massive MIMO technology requires an antenna system with 64T64R or 128T128R capability and multiple sets of radio frequency units. In 5G age, the investment in base station antenna construction will far exceed that for 4G, about ¥50 billion in total.

5G will support the scale-up of industrial innovation and drive investment in ultra-high-definition, virtual reality and other industrial chain. On the one hand, 5G consists of both high frequency and low frequency spectrums. In the early stage of commercial deployment, the emphasis lies in the construction of macro base stations, a low frequency carrier. In the mid-to-late stage, the emphasis shifts to small base stations in order to achieve seamless deep coverage of high frequency network.

Small-and-medium-sized equipment manufacturers and IT device suppliers actively pour into small base station market. It is estimated that global indoor small base station market will reach $1.8 billion in 2021. On the other hand, the integration of 5G with various sectors such as transportation, healthcare, industry, culture and entertainment gives rise to innovative IT products and services as well as various 5G industry applications, such as 4 K/8 K video and virtual reality, which will reshape the development mode of traditional industries.

3.5 Rapid Progress of Industrial Digitalization

New-generation information technology leaps forward and deepens in the penetration into traditional industries. Real economy advances quickly in digitalization, networking and intelligence. Digital technology enormously improves production and operation efficiency. Industrial Internet, Fintech, and artificial intelligence applications become hot fields and achieve large-scale growth.

3.5.1 "Smart+" Becomes a New Form of Economic Development

The integration and penetration of artificial intelligence into traditional industries speeds up with unparalleled incremental effect, multiplier effect and technological spillover effect that traditional industries cannot match. This reshapes a data-driven smart economy featured by human–machine collaboration, cross-discipline integration, and sharing and co-development. The in-depth integration of AI technology with agriculture, manufacturing, finance, education, healthcare, retail and other fields not only promotes large-scale application of artificial intelligence, but also boosts the intelligent level of other industries.

In the integration of artificial intelligence with real economy, as Markets and Markets predicts, from 2020 to 2022, the compound annual growth rate (CAGR) of artificial intelligence in global Fintech market will exceed 40%; from 2020 to 2023, CAGR of global smart retail market and CAGR of artificial intelligence in education will reach 24% and 47% respectively. By 2020, China's artificial intelligence industry will exceed ¥160 billion and promote the scale of related industries to over ¥1 trillion, playing an increasingly vital role in economic and social development.[2]

[2]https://www.ce.cn/cysc/tech/gd2012/201909/06/t20190906_33097412.shtml.

3.5.2 *Industrial Internet Walking into Reality*

3.5.2.1 The Pattern of "Three Pillars" Taking Shape in Global Industrial Internet Platform

Global industrial Internet platform develops quickly. Advanced information technologies such as cloud computing, big data, and artificial intelligence lead to the wide use of smart applications such as centralized monitoring, forecast operation and maintenance and quality optimization in manufacturing industry, and result in the rapid growth of industrial Internet platform market. According to the data released by Markets and Markets, a research institute, global industrial Internet platform market was $2.57 billion in 2017 and $3.27 billion in 2018. It is estimated that by 2023, the market scale will reach $13.82 billion in total, with an expected CAGR as high as 33.4%. In industrial Internet platform market, the equipment management platform sector develops well and accounts for the largest market share presently.

The United States, European Union and Asian countries are major contributors to global industrial Internet platform market. Owing to the leading role played by pioneers such as GE, PTC, Rockwell, IBM, and Microsoft, as well as the active innovation in cutting-edge technologies, the United States displays prominent advantages in platform development and predictably maintains the dominant position in the market for a while. As European industrial giants, such as Siemens, ABB, Bosch, Schneider, and SAP, keep increasing their investment, Europe draws on its globally advanced manufacturing base and makes rapid progress with platform development, becoming a major competitor of the United States. In addition, the growing need for industrialization in emerging economies, such as mainland China and India, contributes to the continued development of industrial Internet platform in the Asia–Pacific. Naturally, Asia achieves the fastest growth and probably becomes the largest market in the future. See Fig. 3.5 for scale of global industrial Internet platform market.

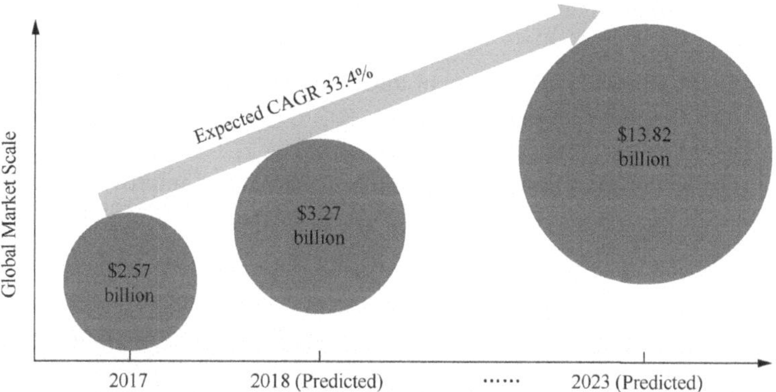

Fig. 3.5 Scale of global industrial internet platform market. *Data Source* Markets and markets

3.5.2.2 More Active Engagement of Industrial Giants and Start-Ups

1. Leading Industrial Giants Take Active Attitude Towards the Deployment of Industrial Internet Platform.

   (1) More attention to digital business. In the fiscal year 2018, Siemens realized the highest return on equity (ROE) at digital firm section, up to 20%, with a year-on-year growth of 1.5%. In its strategy *Vision 2020* + , Siemens announced that it would include digital industry as one of the three major operating fields in the future.

   (2) Sustained improvement in industrial service level. Microsoft Azure IoT platform constantly diversifies the functions of remote device monitoring, predictive maintenance, firm networking and visualization and provides Rolls-Royce (RR), a British corporation, with remote engine operation and maintenance solutions through data collection and analysis.

   (3) Active adjustment of development strategies. General Electric (GE) reconstructs its digital group business as an independently-run company centered on Predix Platform and industrial software to maintain and promote its leading role in digital industry.

2. Innovative Companies in Industrial Internet Platform Play More Active Roles.

   (1) More technological innovation companies emerge. Particle, a company in San Francisco, the United States, launches industrial Internet hardware, software and connectivity platform to help companies track and manage valuable assets, which serves 8,500 companies around the world. MachineMetrics, a start-up in Northampton, provides real-time analysis software, uses machine learning algorithms to analyze data of CNC machine tools and gives advice on operation and maintenance.

   (2) Excellent start-ups are favored by the capital market. Uptake, founded in 2014, has received over $250 million in financing within four years and boosted its market value to $2.3 billion. FogHorn, an edge intelligence software provider, has raised $47.5 million in total.

3.5.2.3 Technological Advancement and Wider Application Promote and Deepen Platform

1. Platform Companies Focus on Core Service Needs and Improve Technological Capabilities.

   (1) Paying attention to industrial real-time application and launching edge solutions. For example, Microsoft launches Azure IoT Edge, an open source platform, and moves cloud-based analysis and business logic to the Edge. Based on Azure IoT Edge, Microsoft and Qualcomm jointly release a new visual AI solution on the Edge.

(2) Improving platform's application and development efficiency and focusing on technologies such as micro-service and low-code development. For instance, Siemens will introduce Mendix low-code technology into its MindSphere platform, which can cut down the time to develop and deploy industrial applications by 90%.
(3) Supporting user's in-depth data mining and consolidating data analysis capability. For example, PTC launches Analytics Manager, a big data analysis and management tool, which can integrate external analysis tools and models into ThingWorx platform.
(4) Enhancing supply capacity of industrial solutions and integrating industrial technology and knowledge. For example, with the help of EcoStruxure platform, Schneider has gathered industrial knowledge from more than 4,000 industrial system integrators and developed innovative applications and solutions for various industries.

2. Platform Construction and Application Advance in a Coordinated Way.

In platform construction, significant progress has been made in capacity expansion and commercial promotion in leading platforms. With substantial growth in assets managed, solutions provided, industrial customers attracted and developers registered on the platform. ThingWorx, PTC's platform, currently hosts over 600 industrial apps used by over 1000 customers every week, and works with more than 380 partners. EcoStruxure platform, deployed at over 480,000 installation sites worldwide, wins support from over 20,000 developers and system integrators and manages over 1.6 million pieces of assets. ABB's ABB Ability platform now collects more than 210 digital solutions. In platform application, it enters a new stage of broader coverage of and deepening penetration into new industries. Platform's targeted customers extend from traditional fields such as petrochemicals, automobiles and electronics to emerging fields such as food and building materials. King's Hawaiian, a food company, connects its machines to Rockwell's FactoryTalk platform to conduct performance monitoring. This helps to produce an extra of 180,000 lb of bread per day and double the output. Platform-based data analysis deepens, turning from basic real-time status monitoring to in-depth analytical prediction. SAP provides predictive maintenance and service solutions to Kaeser, the world's largest air compression system supplier. It can identify and isolate systematic failures, predict certain malfunctions and helps customers in independent maintenance.

3.5.3 Continued Attention on Fintech Industry

3.5.3.1 Continued Rapid Growth of Fintech Industry

According to a survey by CB INSIGHTS, global investment and financing in Fintech nearly quintupled from 2013 to 2017. In 2018, the number amounted to $111.8 billion, with 2,196 investment cases and a new record high. In particular, two acquisitions

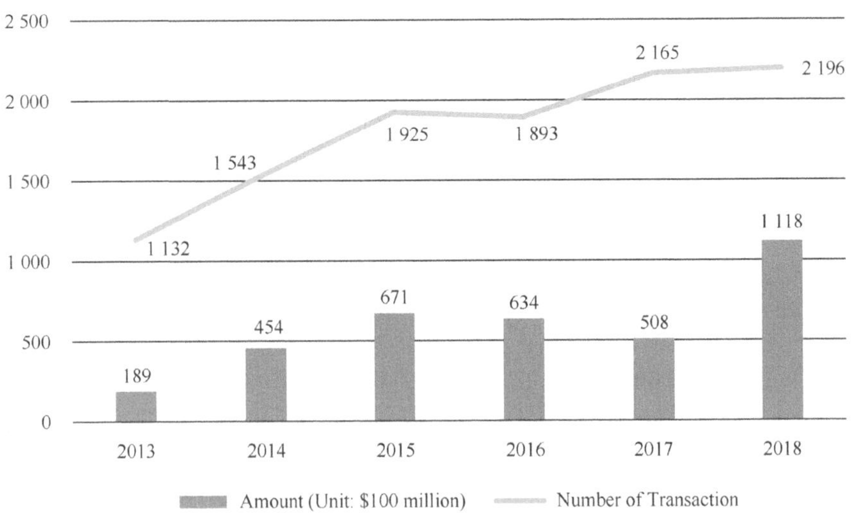

Fig. 3.6 Number and amount of global fintech financing from 2013 to 2018. *Data Source* KPMG

with the largest investment and financing occurred, namely, Ant Financial received $14 billion in C Round of financing and WorldPay completed $12.86 billion worth of merger. See Fig. 3.6 for number and amount of global Fintech financing from 2013 to 2018.

3.5.3.2 North America Takes the Lead and Asia–Pacific Grows Rapidly

In terms of regional development, supported by mature financial service system and strong technological innovation, North America plays a leading role in Fintech. As to the scale of financing, investment and financing in North America, Asia and Europe totaled $65.562 billion from 2014 to 2018, with North America accounting for 53.2%, higher than the sum of Asian and European markets. Asia grows the fastest, its Fintech industry received $6.061 billion in financing in the first three quarters of 2018, nearly 6 times of that in 2014, as shown in Fig. 3.7.

3.5.3.3 Fintech is Most Widely Applied in Payment and Lending Sectors

Among the application fields worldwide, Fintech is most widely used in payment and lending sectors. According to *World's Top 100 Fintech Companies in 2018* ranked by KPMG, payment companies dominate the list. Among them, 34 are in payment sector, 22 are in lending sector, far ahead of other sectors on this list. See Fig. 3.8 for business fields of World's Top 100 Fintech Companies in 2018.

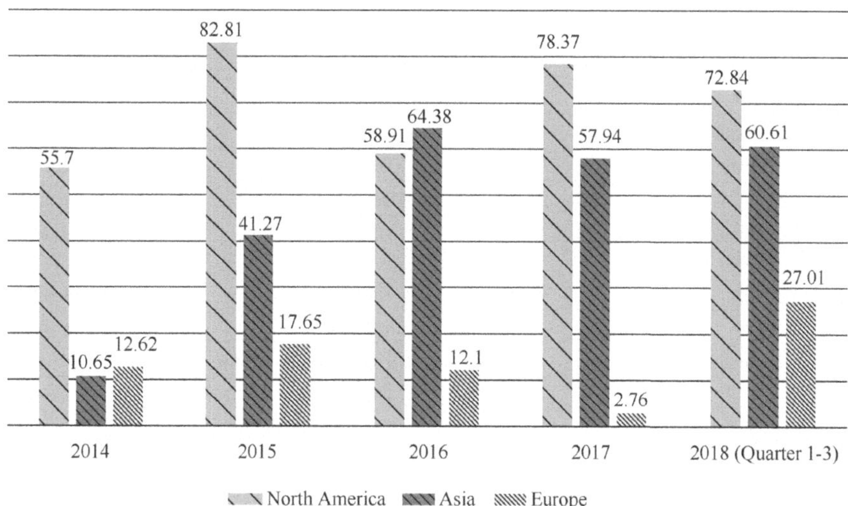

Fig. 3.7 Total financing in North America, Asia and Europe. *Data Source* CB INSIGHTS

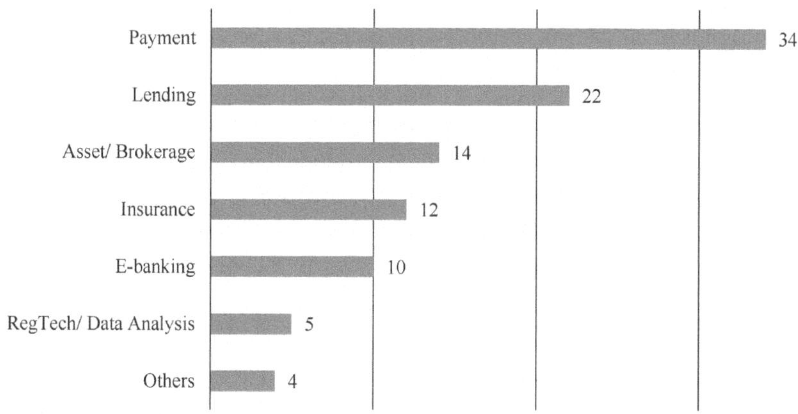

Fig. 3.8 Business fields of world's top 100 Fintech companies in 2018. *Data Source* KPMG

An analysis of top ten financing projects in global Fintech industry in 2018 (see Table 3.2) shows that, payment/transaction sectors account for half of the list, with continued interest in investment and financing in related fields.

3.5.3.4 Innovative Products Bring Challenges to Financial Regulation

In June 2019, 27 institutions headed by Facebook released *Libra White Paper*, planning to launch Libra, a virtual currency. The act cast great influence. Modeling on

Table 3.2 Top ten financing projects in global fintech sector in 2018

No	Company	Amount/$100 million	Region	Business fields
1	Refinitiv	170	The United States	B2B
2	Ant Financial	140	China	Payment/transaction
3	WorldPay	129	The United Kingdom	Payment/transaction
4	Nets	55	Denmark	Payment/transaction
5	Blackhawk Network Holdings	35	The United States	Payment/transaction
6	VeriFone	34	The United States	Payment/transaction
7	iZettle	22	Sweden	Payment/transaction
8	Fidessa Group	21	The United Kingdom	B2B
9	Ipero	19	The United States	B2B
10	IRIS Software Group	17	The United Kingdom	B2B

Facebook, firms are lining up to put virtual currency on their agenda. Featured by stability, low inflation, global acceptance and convertibility, Libra mainly focuses on payment and cross-border remittance, and functions like existing currencies as a unit of account, a medium of exchange, and a store of value, it will greatly impact the existing financial system, currency system and reserve system, and call for new requirements on global financial regulation.

At present, the rapid development of digital economy has not only opened new space with new opportunities, but also created many new challenges. Facing the future, countries ought to actively seize opportunities produced by the new round of technological revolution and industrial transformation, strengthen international cooperation, and further exert their comparative advantages. They should jointly optimize global economic resource allocation, improve global industrial layout, and expand global markets for the benefits of all parties, to fully tap into the potential of digital economy and share the development fruit thereon.

Chapter 4
Development of World Digital Government

4.1 Outline

Under the circumstance of informatization, governments of various countries actively adapt to the development of information technology and figure out the new laws on governance. By reforming idea, coordinating planning, forecasting layout, and strengthening innovation, they quickly transform traditional governance with information technology and vigorously promote the construction of digital government. And digital government becomes a vital support in practicing modern administrative concept, enhancing public administration capability, and strengthening national competitiveness.

The construction of digital government around the world extends to more fields with wider coverage and deeper level. Countries establish high-level, cross-field and cross-departmental coordination mechanisms and founded special agencies to improve the operation system of digital government and accelerate the digital transformation of government. They build cloud platform, strengthen collaboration and integration, and quicken the digitalization of government data resources and infrastructure to make full use of digital technology in daily analysis, decision-making and regulation for more integrative, comprehensive, proactive and precise government services. They construct network platform, enhance collaboration with companies and social organizations, provide open and cooperative public service, and promote digital, mobile and intelligent public service so as to meet the increasing demand for public service. They implement digital identity strategy, vigorously develop digital partnership with all sectors of society, and actively build development environment for digital government. Digital government plays an increasingly important role in promoting economic and social development, improving governance efficiency and enhancing people's welfare.

Chinese Academy of Cyberspace Studies, *World Internet Development Report 2019*, https://doi.org/10.1007/978-981-33-6938-2_4

4.2 Significantly Stronger Support of Information Infrastructure

Information infrastructure serves as foundation for the construction and development of digital government. In recent years, countries around the world seize opportunities of new-generation information technology, optimize network access, promote the construction and application of government cloud, upgrade smart city, and boost the ubiquitous, cloud-based and intelligent development of the infrastructure in digital government.

4.2.1 More Convenient and Compatible Network Access

4.2.1.1 Speeding up the Construction of Network Infrastructure

On the one hand, countries have increased investment in more network infrastructure and broader access. For example, in order to help people living in a scattered way in remote areas gain high-speed network access, Canadian government invested C$85 million in 2019 to develop low-orbit satellites with Telesat, a world-renowned satellite communications corporation, and provided rural and remote areas with high-speed network access. On the other hand, governments set up network access points by themselves or in collaboration with communities, schools, banks and other institutions in densely populated areas. For example, Ministry of Communications of Ghana built network base stations in four western communities in 2018 to provide 1 km-diameter Wi-Fi network and convenient network access to remote areas with relatively dense population[1].

4.2.1.2 Starting Multiple Access and Full Integration of Government Services

In order to facilitate easier and broader access to government services, some countries have developed integrated government apps and integrated self-service terminals which enable users to handle business related to multiple departments with "one-stop access and one-website service". For example, based on the wide use of smart phones, Singapore government provides citizens with access to e-services through mobile applications, and gives the public and companies better access to online government resources.[2] Ministry of Electronics and Information Technology of India developed UMANG, a government service platform (see Fig. 4.1). As of August 2019, UMANG

---

[1]The Ghana Web, 2018, Communications Minister to launch Smart Communities Project. https://www.ghanaweb.com/GhanaHomePage/NewsArchive/Communications-Minister-to-launch-Smart-Communities-Project-62363329.

[2]Data Source: *United Nations E-Government Survey 2018.*

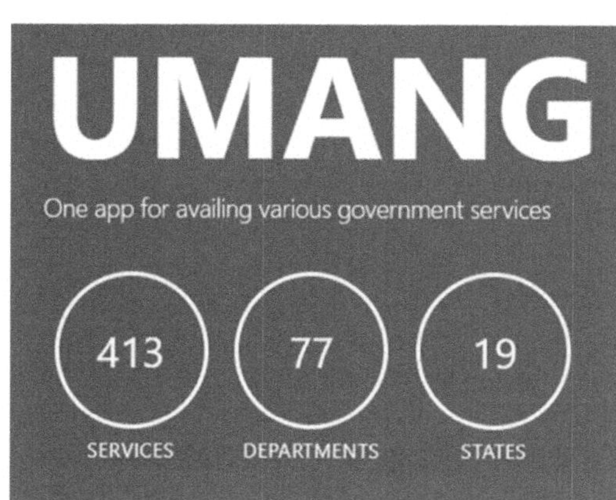

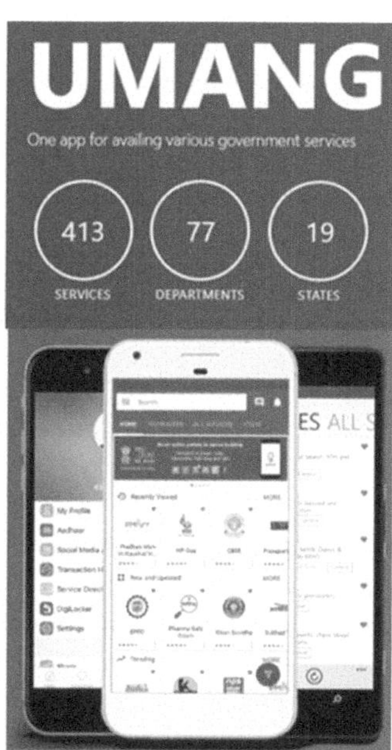

Fig. 4.1 India's UMANG government service platform

integrated 19 Pradesh, 77 departments and 413 services, which facilitated the public's affairs. Meanwhile, it supports several login approaches, and the public can obtain relevant services on their mobile phones.[3]

4.2.2 Cloud Computing Boosting E-Government Development Efficiency

Some countries have developed, co-developed or leased cloud services to construct a digital environment to run e-government. Correspondingly, operation and maintenance cost decreases, government's ability to store and process data improves, computing capability quickens, and equipment maintenance and management become more convenient and efficient.

4.2.2.1 Government Cloud as a Key Direction of Digital Government

For example, U.S. General Services Administration (GSA) of establishes a government cloud platform (Cloud.gov) to handle data from various federal agencies

[3]Data Source: https://web.umang.gov.in/web/#/.

and transfer original service systems onto the cloud. Based on the platform, the U.S. Federal Election Commission (FEC) redesigns the federal election website (FEC.gov), deposits it on the cloud platform, and transfers various data and applications onto the cloud for management, saving 85% in cost every year. In 2018, Myanmar started to build e-Government Integrated Data Center (e-GIDC), based on cloud computing, which provides citizens with services such as tax collection, electronic ID card, and electronic visa, and ensures better communication between the government, firms and the public.

4.2.2.2 Tech Giants Launched Government Cloud Platforms in Succession

For example, there are Amazon's AWS GovCloud, Microsoft's Azure Government, and Huawei's eGovernment CLOUD. In order to guarantee the legitimacy of cloud platform and the security of data, government clouds are usually constructed and operated separately from the private clouds for enterprises. Amazon has developed two government cloud centers for departments across the eastern and western United States. By September 2019, Amazon's government clouds had served over 5000 governments and agencies, including the U.S. federal government, U.S. Department of Defense and other ministries with extremely high requirements on operational performance and security.[4]

4.2.3 Wide Upgrading of Smart City Infrastructure

Cities are the main space of public life and business operation. In the process of digital government, most countries attach importance to boosting the network, digital and intelligent capability of urban infrastructure and fostering good environment in order to improve public service, optimize urban governance, and enhance government management.

4.2.3.1 The Construction of Digital Infrastructure as Key Basic Project of Governments Around the World

For example, by building a high-precision 3D digital model of the city with the simulation of urban infrastructure including water supply, drainage, gas, heating, power and fire protection, governments can monitor the city's utility network in real time and achieve precision management. In terms of urban utility management network, London, New York and other cities have built digital models of pipelines

[4]Amazon China, cloud trusted by government, visit https://aws.amazon.com/cn/government-education/government/.

and spatial information based on Geo-Information System (GIS) and intelligently monitored the flow, temperature, pressure and other working status of the network through IoT technology, which reports accidents to inspectors at the earliest possible time for fast response.

4.2.3.2 Introduction of New Concepts like Digital Twins into Urban Construction

Digital twin city refers to an identical virtual city in the cyberspace recreated by digitizing the physical elements of the real city, where the real city (in physical dimension) and the digital city (in information dimension) co-exist and the reality and the virtual combine. Rennes, France, establishes a 3D digital model of the city for urban planning, decision-making, management and citizen services. Toronto, Canada, plans to build high-tech communities in some parts of its coastal areas, and use various sensors to collect information on traffic, noise, air quality, energy consumption, transportation, and garbage disposal, so as to master the rule and optimize the management of urban operation. When it was set up, China's Xiong'an New District practiced the planning and construction of digital Xiong'an together with the real-sense Xiong'an, to actualize the real-time control of smart application services across the area and build a world-leading digital city with deep learning capability.[5]

4.3 Maturity of Digital Government System and Mechanism

The construction of digital government around the world extends to more fields with wider coverage and deeper level, and continues to achieve important results in policy system, organizing mechanism, leading capability and professional collaboration.

4.3.1 Progress in Policy System of Digital Government

In face of new opportunities and challenges that come with new-generation information technology, countries have formulated various strategic plans, policies and measures to accelerate the construction and development of digital government in recent years.

In regard of overall planning and deployment, developed countries have raced to promulgate strategic plans for the development of digital government. The United Kingdom takes the lead in the construction of digital government. In 2017, based on*UK Digital Strategy* and *Government as a Platform*, the United Kingdom released

[5]Data Source: *Digital Twin City Research Report (2018)*, China Academy of Information and Communications Technology.

Government Transformation Strategy (2017–2020). It proposed to facilitate overall transformation of interdepartmental government services, promote digital technology, optimize business tools, processes and management models, make better use of data, and develop sharing platforms, components and reusable business function to quicken the digital transformation of government.[6] The United States released *Digital Government: Building a 21st Century Platform to Better Serve the American People*. By opening government data and promoting application innovation, it aimed to continuously improve the quality of government services and ensure its leading role in the digital world.

In terms of development policies for special fields, developed countries such as the United States, the United Kingdom and Australia have formulated quite a few on relevant areas, such as big data, artificial intelligence and open data, to support the construction and development of digital government. Take the United States as an example. From 2017 to 2018, the United States passed *Artificial Intelligence Innovation Team Act*, *Future of Artificial Intelligence Act of 2017*,[7] *Artificial Intelligence Employment Act* and other regulations. In 2019, the United States updated *National Artificial Intelligence Research and Development Strategic Plan*, to support full-spectrum technological R&D and innovation and ensure America's leading role in the field of artificial intelligence. In 2019, the United States issued *Open Government Data Act*, demanding that government should establish a complete data list and make regular updates, set up chief data officer and managing committee, and develop a report and evaluation system for open government data, which provided strong support for opening, sharing and application of government data. In addition, the United States also formulated *Federal Data Strategy* to push the government to make better and more efficient use of their data assets and improve its data governance quality.

4.3.2 Growing Overall Planning and Coordination Efforts

As governments push on their digital transformation, problems with traditional governance such as overlapping management, crossover function, inconsistent power and responsibility and poor efficiency grow increasingly prominent, which urgently calls for stronger overall planning and coordination between departments and along the managerial hierarchy.

[6] Zhang Xiao, Bao Jing. "Government as a Platform: Research on Digital Transformation Strategy of the UK Government and Its Enlightenment [J]. *Chinese Public Administration*, 2018.

[7] Data Source: *Global Artificial Intelligence Strategy and Policy Outlook (2019)*, by China Academy of Information and Communications Technology, August 2019.

4.3.2.1 Building Cross-Departmental Overall Planning and Coordination System

Building digital government is a systematic project, which requires stronger systematic integration and coordination to effectively break the governance barrier between different departments, hierarchy and regions. In 2018, Australia established Australian Data and Digital Council (ADC) to integrate national government data, coordinate digital government work, and guide and manage the digital transformation in various states and areas. In 2019, ADC released an overview of Australia's national data and digital government plan, with 93 projects at federal and state levels on the list. China has established a national e-government coordination system, in which Office of the Central Cyberspace Affairs Commission leads and National Development and Reform Commission takes part in, to coordinate the development of e-government nationwide.

4.3.2.2 Strengthening Information Coordination Responsibilities of Government Officials

In face of the challenges in information age, government departments and staff must actively adapt, integrate information technology with government service in a deeper way, and fulfill their duties using Internet and information technology. For example, many government departments of the United States have set up chief information officers (CIO) and chief data officers (CDO) who are responsible for formulating informatization plans and strategies, tracking project implementation, developing and utilizing government information resources, and improving department's information capability. This bridges information technology with government services and coordinates the digital development within departments.

4.3.3 Significantly Improved Professional Management and Coordination Capability

Based on strengthened overall coordination, most governments have adopted measures, such as setting up special data management and coordination institutions and stepping up professional collaboration, to actively promote cross-hierarchy, cross-departmental and cross-regional digital coordination within governments.

4.3.3.1 Establishing Specialized Agency to Promote the Digital Transformation of Government

For instance, British Cabinet Office sets up the Government Digital Service (GDS), and Swedish government establishes Agency for Digital Government (DIGG). Their functions include organizing and implementing digital government work and promoting the construction and application of government service systems and databases. The U.S. government sets up Digital Government Research Center, and proposes to improve the quality of digital government through evaluation-facilitated construction.[8] Australia establishes Victorian Centre for Data Insights (VCDI) to collect public service data, cooperate with other government departments and agencies on data analysis projects, and to promote data-use cooperation between Australia and other countries. In 2018, Australia Digital Transformation Agency (DTA) launched the digital community initiative, which contributed to solve community public problems through inter-governmental cooperation, specifically online forums, regular meetings and other online and offline activities.

4.3.3.2 Growing Digital Cooperation Between Countries and Regions

Denmark, Finland, Iceland, Norway and Sweden establish the Nordic Council of Minister for Digitalization 2017–2020 to coordinate and promote the strategic deployment of digitalization in the Nordic area. In 2018, representatives from eight Nordic and Baltic countries, Sweden, Denmark, Finland, Norway, Iceland, Estonia, Latvia and Lithuania, signed *Declaration of Cooperation on Artificial Intelligence* to promote better use of artificial intelligence by government agencies.

4.4 Significantly Improved Information Application in Government Services

As digital government advances, more and more countries keep digital documents and records, and more agencies rely on digital government information for cross-department coordination. At present, the management, development and utilization of government information in various countries are moving towards digitization, openness, sharing, and information as a resource.

---

[8] *Enlightenment and Lessons from Global Digital Government Transformation* issued by Department of Big Data Development of the State Information Center, visit https://www.sic.gov.cn/News/612/9842.htm.

4.4.1 Speeding Up Digitalization of Information Resources

The digital level of information resources plays a crucial role in achieving synergy in the construction of digital government. Now, governments around the world still run on hard copy documents, and many agencies depend on traditional medium to store and manage government information, which limits the effectiveness of digital government. However, some countries speed up the digitalization of information resources and deepen the application of digital technologies in public services.

4.4.1.1 Digitalization as an Increasingly Important Direction for Creating and Storing Information Resources

A typical use case is the sharing of electronic health records within the healthcare system. Italy invests 750 million euros to set up electronic health records and digitize medical prescriptions, develop telemedicine, and promote online appointments for all its citizens, in order to optimize healthcare resources and reduce waiting time for treatment. Austria designs a digital healthcare system for all citizens to achieve better coordination between its financial institutions, healthcare institutions and stakeholders and simplify procedures for the settlement of medical expenses. Australia focuses its digital economy strategy on electronic records and telemedicine, and plans to increase the sharing rate of personal electronic health records accessible to the elderly, pregnant women, infants and patients with chronic conditions to 90% by 2020, so as to gradually promote remote healthcare services such as telemedicine insurance scheme, general practitioner video consultation hotline, and pregnancy and newborn help hotline.

4.4.1.2 Widening Coverage of Digital Information Resources

With the wide use of digital audio–video devices, scanners, storage media and other equipment, governments step up the digitalization of existing records and archived resources, and carry out more services online to form new digital information resources. UK National Archives stated in *Digital Strategy* in 2017 that given the urgent need of digital transformation of government records, archiving departments ought to upgrade their digital storage capabilities and push for revolutionary digital archive. The website of UK National Archives is shown in Fig. 4.2. In 2018, German government issued *Federal Government's Implementation Strategy for Promoting Digital Transformation*,[9] which confirmed the management of government imagery resources as the basic digital work.

[9]Source: https://germandigitaltechnologies.de/national-strategies/

Fig. 4.2 Website of UK national archives

4.4.2 *Speeding Up the Construction of Data Sharing Platform*

In order to promote digital collaboration among government agencies, countries around the world are vigorously building government data exchange and sharing platforms to provide support for the efficient circulation and timely transfer of government information. Since most data sharing platforms require connecting government agencies at different hierarchy across different regions with business systems, these projects are harder and cost more. Therefore, some countries have adopted flexible and diverse methods to encourage the development of shared government data interfaces and to accelerate the construction of government data sharing platforms.

Presently, there are two main approaches to develop government data interface in various countries:

(1) Data aggregation-based approach. Departments provide sharable data resources, and the platform integrates, aggregates and processes them appropriately and then shares them when needed. American and British government agencies adopt the approach to disclosure and share data resources.
(2) Data interface-based approach. Departments set up their data interfaces, and the platform manages all these interfaces and provides data sharing services through them. For example, the government of the State of Victoria, Australia builds a data interface factory and gateways to help agencies develop data interfaces and manage them in a unified way. The state also formulates best-practice design standards for data interfaces and establishes a portal website to share government data, which shapes a favorable environment for cross-department and cross-region government data sharing.

4.4.3 Continuous Effect Generated by Data Disclosure

Data disclosure is an important measure to promote information resources innovation, build transparent government, and maximize the benefit of information for the public. In recent years, developed countries have actively promoted the openness of government data, and the value of data resources has emerged.

4.4.3.1 Countries Have Formulated Disclosing Public Data Resources Plans

Developed countries such as the United States, the United Kingdom, Australia, Canada, and New Zealand have announced disclosing public data plans. And to disclose government data turns to a global trend and common view. The United Kingdom is a pioneer in disclosing government data. At present, the scope of data disclosure extends to many fields on people's life, such as social welfare, law, taxation, transportation, education, employment and immigration, and contributes to the promotion of an open transparent government and social innovation and entrepreneurship. The United States actively promotes government data disclosure. As of August 2019, the federal government's open data portal (DataGov) has provided 236,391 data sets, covering 14 fields including agriculture, climate, consumption, education, and energy. The data are listed with detailed descriptions of title, use, reference, meta-data and other attributes, and can be downloaded in different formats.

4.4.3.2 Data Disclosure Generates Its Value

Government's data disclosure helps develop and mine the value of data resources and provide new ideas, methods and approaches to solve various problems encountered in social and economic development. For example, New York uses open government data to improve management and encourage entrepreneurship and innovation among the public. New York continually holds NYC BigApps competition to encourage the use of open government data for application development and boost people's participation in big data application and urban management. Winners are not only generously rewarded in cash, but also given the opportunity to promote their works on New York network platform. Data sharing and disclosure contributes to the integration and innovation of government data and social data and strongly promotes social development and reform.

4.5 Continuous Improvement of Public Service Capability

As information technology advances, mobile Internet grows more prevalent in recent years. New-generation information technologies such as artificial intelligence and big data are widely applied in government public services, which boost government's ability to serve the public.

4.5.1 Mobile Internet Facilitates More Accessible Public Services

With a growing use of smart phones, "available and accessible anytime anywhere" has become a new requirement for government public services. Some countries have upgraded mobile websites and developed government applications, pushing the mobile and ubiquitous transformation of public services.

4.5.1.1 Strengthening Public Services on Mobile End

Many countries have picked some of the mostly-used public services and kept improving their mobile-end applications, in order to deliver highly-demanded public services on the mobile end. For example, the British government selects 25 mostly-used public services including land registration, passports and visas, education subsidies and vehicle management, and integrates government information and services from 25 government agencies and 405 institutions onto the GOV.UK website. Based on massive visiting from smart phones, the government continues to optimize GOV.UK for simpler and easier access on the mobile end and better user experience.

4.5.1.2 Vigorously Developing Mobile Government Applications

In recent years, there has been an emergence of various government service applications for specialized use. As of August 2019, the U.S. federal government has developed 93 iOS applications, 72 Android applications, and 18 mobile government websites. It has also established a "Federal Government Mobile App Directory" to help the public find official App.[10] China has also witnessed a surge of mobile government applications. For example, Zhejiang Province develops "Zheliban App", a government app that provides consolidated services across the province, including hospital registration, traffic violation processing, housing ownership certificate, tax payment certificate, social security and provident fund inquiry, college entrance

[10]Data Source: https://www.usa.gov/mobile-apps.

examination result and admission inquiry, subway ticket purchasing, parking space finding, public restroom locating, and other services easy to access on the mobile end.

4.5.2 Artificial Intelligence Improves the Precision of Public Services

In recent years, some countries have actively explored the application of artificial intelligence in public sectors to continuously improve the precision, response speed and quality of public services.

4.5.2.1 Improving the Efficiency of Public Services

So far, more than 20 countries and regions have formulated artificial intelligence strategies to promote its application in the public and private sectors. In 2019, the U.S. federal government sped up the development and application of artificial intelligence and automation, and pushed the use of emerging technologies in public services across various agencies. For example, in order to process large numbers of aid and employment applications efficiently, Department of Social Services of New York works with IBM, using Watson, an artificial intelligence platform, to determine the qualification of applicants. Specifically, based on the Watson's machine learning and big data analysis capabilities, the department reviews the applicants' social welfare, income and subsidies, business insurance and reason for application against application requirements quickly, thereby greatly improving processing efficiency and the quality of public services.

4.5.2.2 Improving the Quality of Public Services

Playing a leading role in e-government, Singapore attaches great importance to the application of AI technology in the public sector. In Smart Nation 2025 Strategy, artificial intelligence and big data are listed as key technologies to improve public services and optimize urban management. In order to improve public inquiry services, Singapore has developed a virtual assistant "Ask Jamie" with Microsoft. Based on past public inquiries and machine learning capability, "Ask Jamie" can respond to public consultation quickly and are deployed to more than 70 agencies and institutions.

4.5.3 Big Data Improves the Precision of Public Services

Governments have collected massive data in the line of duty. These data resources, if mined and analyzed fully with big data technology, can contribute to the provision of precise and individualized public services and add more scientific elements in government decision-making.

4.5.3.1 Promoting Targeted Supply of Public Services

Some countries are actively deploying big data technology to predict the needs of the public and companies and explore the possibility to offer services ahead of the occasion. For example, in 2019, Australia upgraded the virtual assistant in social security service, "Virtual Assistants" App. By analyzing basic information of applicants, the App can proactively send reminders and provide services. After getting user's consent, it can construct a digital portrait and suggest targeted services according to different stages the user is in.[11]

4.5.3.2 Boosting Individualized Public Services

In order to improve maternal and child healthcare, Mexican government has teamed up with UNICEF to pilot "Prospera Digital Experiment Program", which determines the needs of pregnant and nursing women by mass texting, imitating dialogue and analyzing responses, and provides targeted assistance accordingly. By the end of 2018, the program has successfully helped 5000 women.[12] Built on big data-supported government services, China's Nanjing has summarized a list of unique and differentiated service sections in the "My Nanjing" App according to the potential needs from the different life cycles of the public and companies. For example, in the most frequently visited sections, such as social security, provident fund and individual income tax, the App sends targeted messages to citizens, which is well-accepted by the public. By August 2019, the installation of the App on Android end alone exceeded 10 million times.

4.5.3.3 Promoting Government's Scientific Decision-Making

Department of Innovation and Technology (DoIT) of Chicago, the United States, has created a series of intelligent decision-making tools using big data and other technologies. By obtaining, utilizing, and identifying data related to the city's daily

---

[11]Data Source: https://www.computerworld.com.au/article/659288/human-services-expand-chatbot-ranks-pipa/.

[12]Data Source: *United Nations E-Government Survey 2018.*

operation in a timely manner, the tools aim to build a data-based scientific decision-making model.

4.6 Improvement of Environment for Digital Government

The construction of digital government requires a supportive environment, one that allows the government to operate in digital space. In recent years, it is a general practice among governments of various countries to start building their national digital identity systems and establish digital partnership with all sectors of society. And they have achieved remarkable results in shaping the environment for digital government.

4.6.1 Digital Identity System Gradually Established

Digital identity is the "identity card" in digital space. As social informatization advances and digital government service deepens, the construction of an identity system in digital space has become the groundwork of digital transformation. At present, some countries have developed digital identity systems for citizens, companies, and public sectors based on cryptography, security chips, and blockchain technology, and passed laws to ensure the consistency with real identities, which strongly supports the transformation and development of digital government.

4.6.1.1 Construction of Digital Identity System for the Public

In recent years, many countries have issued "digital ID cards" to citizens, making it easier for them to obtain online services. In support of the digital transformation of British government led by its Cabinet, Government Digital Services (GDS) establishes an online identity verification system (GOV.UK Verify) that provides one-stop electronic identity services and offers users secured and fast authentication on government websites. British government plans to expand the coverage of its online identification to 25 million citizens in 2020. It also establishes a government payment system (GOV.UK Pay) and a government notification system (GOV.UK Notify) to provide citizens with electronic ID-based online payment and notification services. Estonia has carried out "e-Residency" ID project. Based on blockchain technology, the project issues "electronic resident" ID after receiving the application for government verification. Estonian government provides various online services to citizens and companies that have obtained "electronic resident" ID, including nationwide online voting, e-signature for contracts, all-step online incorporation that takes only 18 min, and tax reporting that covers 95% of the tax bills. This project accepts applications from all over the world. As of April 2019, 53,719 people from

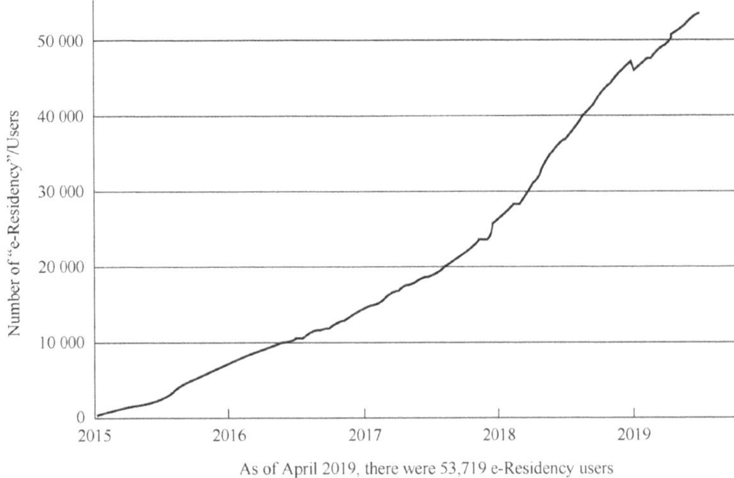

Fig. 4.3 Number of "e-Residency" users in Estonia

over 175 countries applied for Estonian e-Residency, and Estonian government plans to expand that number to 10 million by 2020, as shown in Fig. 4.3.

4.6.1.2 Construction of Corporate Digital Certificate System

Corporate digital identity is fundamental for companies to carry out business activities in digital space. European Union has formulated and implemented *Electronic Identification and Trust Service Regulations* (eIDAS) to help companies that frequently conduct business across borders get access to public services in all EU member states easily and efficiently. In September 2018, eIDAS came into effect. It recognizes the legitimacy of electronic ID of public companies within EU. EU companies can get access to digital services across EU member states by using their electronic ID, which relieves them of the burden of having to submit information and materials to different government agencies several times and saves a lot of money for both companies and governments. The United States uses the existing employer identification number (EIN, also known as Federal Tax Identification Number) for corporate digital identity registration services with online IRS application channels for both domestic and foreign companies. At the end of 2018, the State Administration for Market Regulation of China issued *Regulations on Electronic Business Licenses (Trial)*, clarifying that electronic business licenses have the same legal effect as hard copy business licenses. In June 2019, Electronic Business License Display System was officially launched. The system reviews business licenses submitted by companies and generates links for the public to inquire about corporate information such as business scope, operating period, and registered capital.

4.6.1.3 Digital Signature for Public Sector

Digital identity for the public sector is mainly used for authentication between public agencies when providing services, which means great significance for making public services more easily accessible. Electronic seals and digital signatures help government agencies with online identification, and contribute to better coordination of network, data and services between different departments. In August 2019, China proposed to "promote the application of electronic seals", to change the practice of relying solely on hard copy government seals for authentication and to effectively cut down the time for setting up businesses. In April 2019, "One Website All Services" Platform developed by China's Shanghai launched electronic seal service to uniformly generate and manage e-seals for legal persons and e-signatures for individuals. The service aims to achieve city-wide recognition of electronically signed documents issued by different government agencies and reduce operation cost for businesses and save citizens a trip to get services.

4.6.2 Further Deepening Government-Enterprise Cooperation

Enterprise holds great technological and capital advantages. At present, more and more enterprises are participating in the construction, management and operation of digital government, and constituting an increasingly important force to promote digital government.

4.6.2.1 More Open Development Model of Digital Government

High software/hardware investment and information system maintenance cost are common problems of the digital transformation of governments around the world. In recent years, more and more governments are working with enterprises to reduce their cost and improve efficiency. Some countries have adopted Public–Private Partnership (PPP) approach for the construction of digital government to achieve a win–win outcome for both sides. For example, Indiana State Government of the United States establishes an integrated online service system for free through Build-Operate-Transfer (BOT). In the BOT method, the enterprise undertakes the responsibility of the development, operation, maintenance and management of the system, charges about 1% of the services on the system to cover its cost, and transfers the system to the government after certain years of operation. In this way, the government can provide better, more efficient and digital services to the public without any prior funding. Build-Own-Operate (BOO) is widely used in the United Kingdom to develop government cloud, where the operator finances, builds and owns cloud infrastructure, and the government pays only for online systems and offline proxy services. This significantly reduces the initial investment and the cost for upgrading and maintenance in infrastructure for the government.

4.6.2.2 Providing Government Information and Services Through Enterprise Platform

Some countries collaborate with enterprises and social organizations to provide public services through social media or other platforms, which enlarges the channel and form of services and better meets the growing need for information and services from the public and enterprises. According to *United Nations E-Government Survey 2018*, the number of countries using social media to post information and provide services has increased from 152 in 2016 to 177 in 2018. The United States, European Union and other countries and regions cooperate with social media or digital payment platforms to provide public services such as utility bill payment and information inquiry. China's online platforms such as Alipay and WeChat integrate and provide a large number of government services and public services, covering social security, transportation, healthcare and environmental protection. As of June 2019, Alipay alone has launched government services in 442 cities.

Chapter 5
Development of World Internet Media

5.1 Outline

Featured by massive information, in-depth sharing, instant interaction, and rapid dissemination, Internet has become an increasingly important channel of communication for people all over the world. Internet media continue to innovate in terms of product form and platform, and constitute a growing force in expanding communication space and promoting learning and exchange between civilizations. The iteration and upgrading of new media, new technologies and new applications has produced profound changes to public opinion ecosystem, media landscape, and communication methods. As Internet media grow more mobile and intelligent every day, they have become the main channel and platform for information dissemination and important carrier to inherit and disseminate human's excellent cultural achievements.

In 2019, global Internet media industries develop steadily with some changes. Most of the world's major Internet media companies are located in the United States, China and other countries and regions. Global social platforms vary greatly in their user coverage, with East Asia, North America and North Europe far ahead of Africa and Central Asia in penetration rate. Social media become an important channel for news with emerging trends such as younger user groups, diversified publishing platforms and private news consumption.

The development of information technology contributes to a series of profound changes in Internet media from the news production end which includes material acquisition and editing, content production, data distribution, and decision-making, to the consumption end which includes content display, user experience and communication effect. Specifically, 5G technology deeply improves user experience, cloud computing provides easier, more flexible and efficient ways for online contents production, and artificial intelligence enormously impacts all the media value chain.

However, it's hard to control the dissemination of false information on social media and instant messaging platforms and curb the spread of terrorism and violent extremist content on Internet. The abuse of deep fake and other technologies has posed new challenges to online content regulation. As large Internet media platforms

Chinese Academy of Cyberspace Studies, *World Internet Development Report 2019*, https://doi.org/10.1007/978-981-33-6938-2_5

further monopolize the market, governments take the problem seriously and tighten their anti-monopoly laws.

5.2 Global Landscape of Internet Media Development

In 2019, global Internet media industries develop steadily with the United States and China taking the lead. There are twelve Internet companies involved in media business that are valued or evaluated over $25 billion, with their products and services covering social media, digital media, content aggregation platform and other sectors. Among them, six companies are headquartered in the United States, five in China, and one in Sweden.[1] See Table 5.1 for major Internet companies involved in media business around the world.

Table 5.1 List of major global Internet companies involved in media business

No.	Company	Country (headquarters)	Main product/business category	Market value/valuation (unit: $1 billion)
1	Microsoft	The United States	LinkedIn, Azur, etc.	1007
2	Amazon	The United States	AWS Cloud Service	888
3	Alphabet Inc.	The United States	Google, YouTube, etc.	741
4	Facebook	The United States	Facebook, Instagram, WhatsApp, etc.	495
5	Alibaba	China	YouTube, UC Browser, Ali Pictures, DingTalk, Alicloud, Xiami Music, etc.	402
6	Tencent	China	QQ, WeChat, Qzone, QQ Music, Tencent Video, etc.	398
7	Netflix	The United States	Internet video ecology, etc.	158
8	ByteDance	China	TouTiao, TikTok, Top Buzz, News Republic, Ixigua, Buzz Video, Vigo Video, etc.	75

(continued)

[1] Integrate Data Source: Bond, Internet Trend 2019, June 11, 2019, https://www.bondcap.com/pdf/Internet_Trends_2019.pdf. Bytedance is China's most successful international internet company, valued at $75 billion, 2019.4.19, https://chinaeconomicreview.com/bytedance-is-chinas-most-successful-international-internet-company-valued-at-75-billion/.

Table 5.1 (continued)

No.	Company	Country (headquarters)	Main product/business category	Market value/valuation (unit: $1 billion)
9	Baidu	China	IQiyi, Baidu Cloud Disk, Baidu Search, Baidu Feed, Baidu Cloud, etc.	38
10	NetEase	China	NetEase, NetEase News, NetEase Cloud Reading, NetEase Cloud Music, NetEase Blog, NetEase Open Class, etc.	33
11	Twitter	The United States	Twitter	29
12	Spotify	Sweden	The world's largest music streaming subscription service	25

5.2.1 Social Platform

Now, global social platforms are characterized by numerous users, great regional disparity and high geographic concentration. In 2019, the total monthly active users of popular social platforms worldwide reached 3534 million, accounting for 46% of the global population. Among them, the total number of users with mobile access was 3463 million.[2]

There is obvious regional disparity regarding the number of users and penetration rate of social platforms across the globe. East Asia, South Asia, Southeast Asia, South America, and North America have lots of social platform users, while Central Africa and Central Asia have the least number of users. East Asia, North America, and North Europe score high on user penetration rate, all exceeding 65% of the total population. While that in Africa and Central Asia yield below 20%, leaving huge room for growth.

In January 2019, in terms of monthly active users on social platforms, East Asia yielded the highest number, over 1100 million, followed by South Asia (449 million), Southeast Asia (402 million), South America (285 million) and North America (255 million). Central Africa and Central Asia were at the bottom on the list, and both yielded 12 million, as shown in Fig. 5.1.

In terms of user penetration rate, East Asia and North America have the highest rate of 70%, followed by North Europe (67%), South America (66%), Central America (62%) and Southeast Asia (61%), and East Africa and Central Africa were under 10%, as shown in Fig. 5.2.

[2]The data is based on the monthly active users of the most popular social platforms of each country. Data Source: Hootsuite, July 28, 2019.

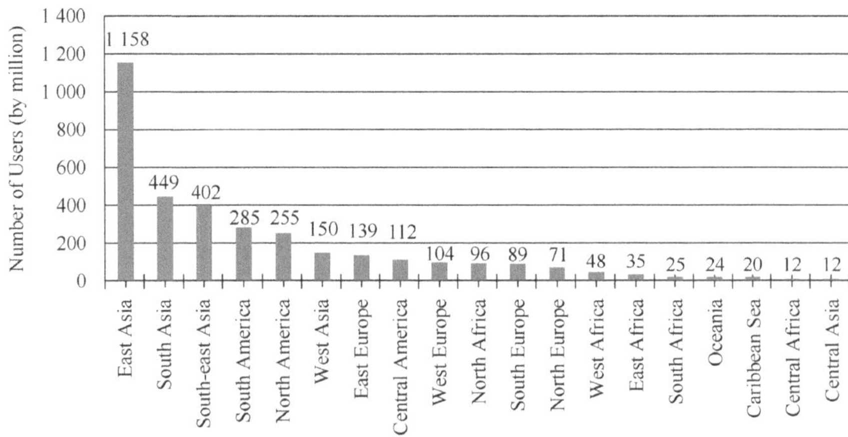

Fig. 5.1 Overview of the use of social platforms by region in 2019. *Data Source Global Digital 2019Reports* released by Hootsuite. *Note* Calculated based on users active on social platforms each month

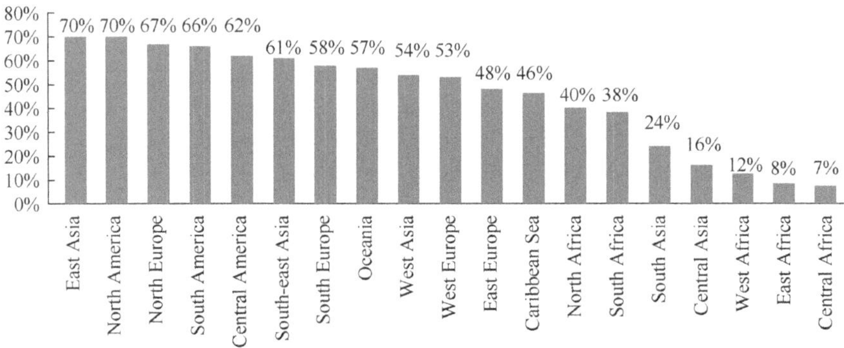

Fig. 5.2 Penetration rate of social platforms in the whole population in 2019. *Data Source Digital 2019 Reports* released by Hootsuite

The world's most popular social platforms are mainly located in the United States and China. According to statistics, in terms of monthly active users, as of July 2019, among the top eighteen popular social platforms in the world, eleven are in the United States, six in China, and one in Japan,[3] as shown in Table 5.2.

The sources of users vary remarkably among the world's major social platforms. Facebook, Instagram and Twitter are better at developing overseas markets, and many of their users are located at diversified range of countries. Among them, Facebook

[3]Data Source: *Digital 2019 Reports* released by Hootsuite Q3 Global Digital Statshot. As of July 15, 2019.

Table 5.2 Monthly active users on popular social platforms

No.	Platform	Company	User/million
1	Facebook	Facebook (The United States)	2375
2	YouTube	Google (The United States)	2000
3	WhatsApp	Facebook (The United States)	1600
4	Facebook Messenger	Facebook (The United States)	1300
5	WeChat	Tencent (China)	1112
6	Instagram	Facebook (The United States)	1000
7	QQ	Tencent (China)	823
8	Qzone	Tencent (China)	572
9	TikTok	ByteDance (China)	500
10	Sina Microblog	Sina (China)	465
11	Reddit	Advance Publications (The United States)	330
12	Twitter	Twitter (The United States)	330
13	Douban	Beijing Douwang Technology (China)	320
14	LinkedIn	Microsoft (The United States)	310
15	Snapchat	Snap Inc. (The United States)	294
16	Pinterest	Pinterest Inc. (The United States)	265
17	Viber	Rakuten (Japan)	260
18	Discord	Discord Inc. (The United States)	250

has far more users in India than in the United States. Most of its users are located in Asia and Americas, and users in Africa and Europe are few,[4] as shown in Fig. 5.3.

Instagram users are mainly scattered across the Americas, Asia and Europe, most of which are located in the United States, Brazil, India, and Indonesia ranked in descending order, as shown in Fig. 5.4.

Twitter users are mainly scattered across the Americas, Asia and Europe, with the most in the United States, followed by Japan, as shown in Fig. 5.5.

Snapchat, Pinterest, Weibo, WeChat, Line, and Kakaotalk have obtained market advantages in their own region. Among them, Snapchat has absolute advantage in the United States, its home market, with fairly equal user presence in France, the United Kingdom, Brazil and other countries, as shown in Fig. 5.6.

Pinterest users are mainly located in Americas and Europe, with the United States on the top of the list, as shown in Fig. 5.7.

The users of Line, Kakaotalk, Weibo and WeChat are concentrated in Asia. Line users are mainly from Japan, accounting for 86%, followed by China's Taiwan, Thailand and Indonesia. Kakaotalk has more than 50 million users worldwide, and over 44 million of them are from South Korea. The users of Weibo and WeChat are mainly in China.

---

[4]Relevant data on the source of user extracted from: Statista 2019 Social Media Worldwide, as of July 2019.

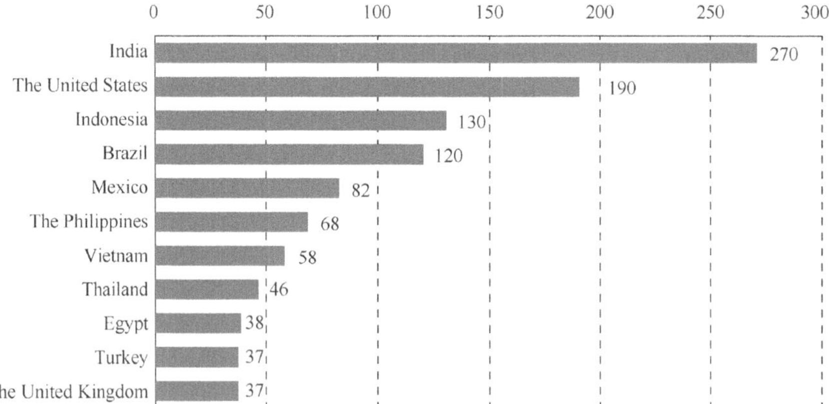

Fig. 5.3 Countries with major Facebook user presence (unit: million)

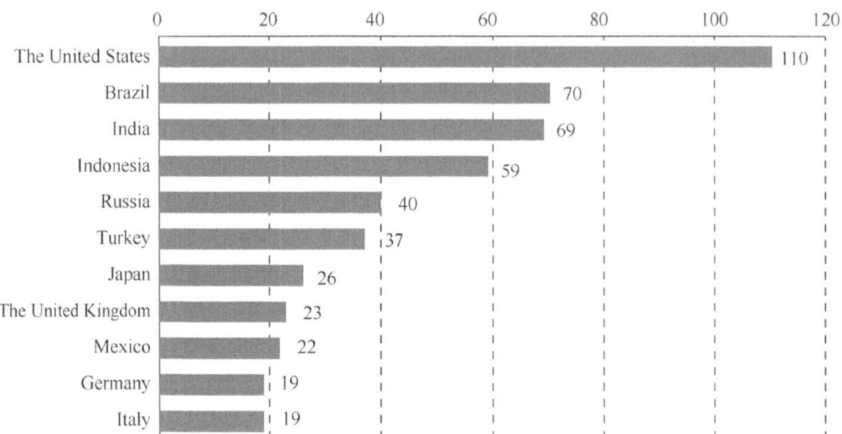

Fig. 5.4 Countries with major Instagram user presence (unit: million)

5.2.2 *Digital News*

5.2.2.1 Digital News Subscription

In 2019, digital news subscribers to many major news outlets continue to grow around the world. According to *Journalism, Media, and Technological Trends and Predictions 2019*, more than half of news outlets believe that digital subscription is the main source of revenue in the future.

In 2019, the number of news outlets that had drew 100,000 or more digital subscribers worldwide reached 27 in total. Geographically, most of them are European and American news outlets. Specifically, eight are in the United States, fifteen

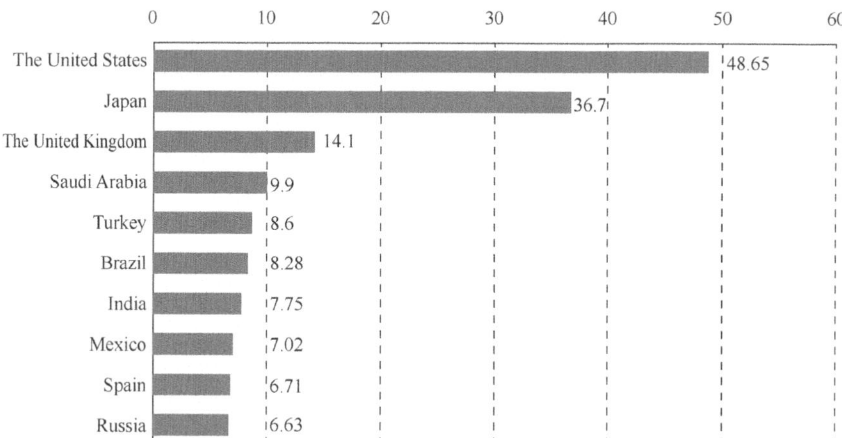

Fig. 5.5 Countries with major Twitter user presence (unit: million)

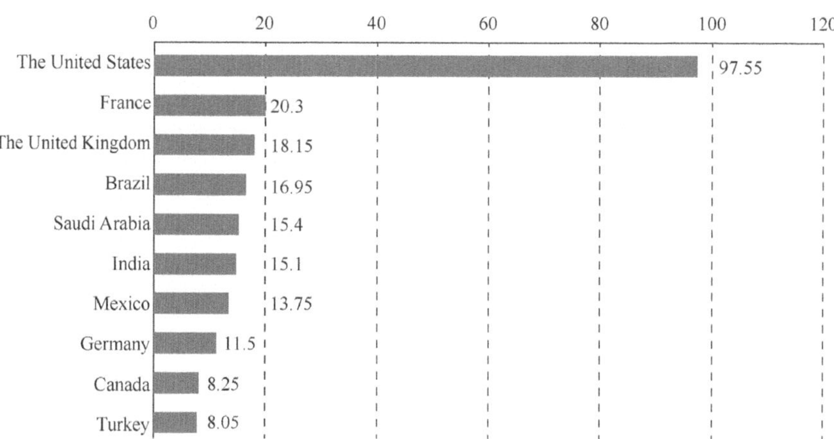

Fig. 5.6 Countries with major Snapchat user presence (unit: million)

in Europe, and each two in Asia and South America. The reasons behind this pattern are as follows:

(1) European news media established a good reputation and brand value in history. Based on reliable and high-quality news content, such media as *The Guardian* of the United Kingdom and *Aftonbladet* of Sweden draw subscribers, which is hard to replicate in other countries.
(2) News media environment affects the desire to subscribe to digital news. When most online news is available for free, there will be fewer digital subscribers, as is the case with China.

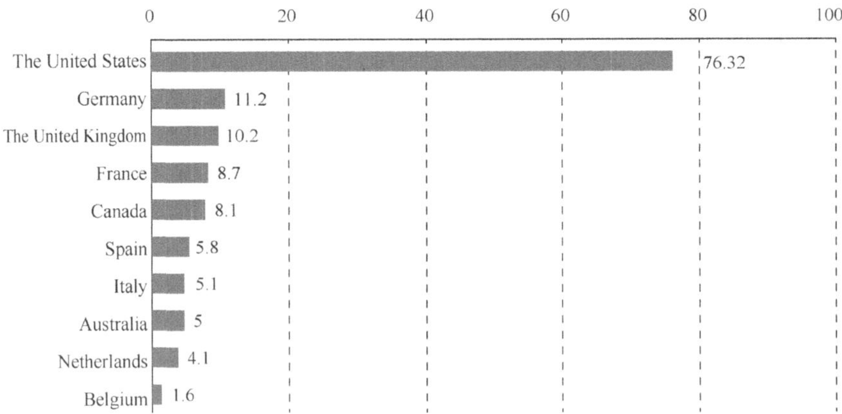

Fig. 5.7 Countries with major Pinterest user presence (unit: million)

(3) People are more willing to pay for news services that cover the professional fields. For example, *Wall Street Journal* of the United States, *Financial Times* of the United Kingdom, *Caixin* of China are all professional financial news media with many digital subscribers. See Fig. 5.8 for digital subscription of major news outlets in 2018 and 2019, and see Table 5.3 for news outlets with over 100,000 digital subscribers worldwide in 2019.

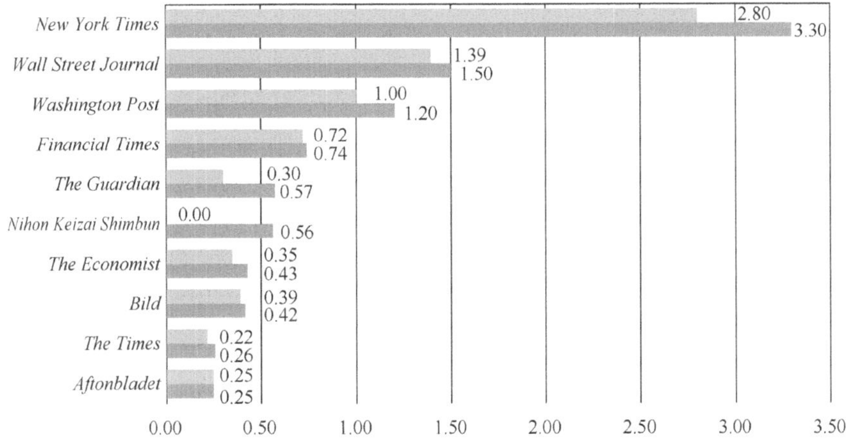

Fig. 5.8 Digital subscription of major news outlets in 2018 and 2019. *Data Source 2019 Global Digital Subscription Snapshot* and *2018 Global Digital Subscription Snapshot* by FIPP and CeleraOne

Table 5.3 News outlets with over 100,000 digital subscribers worldwide in 2019[5]

Rank	Country	Name	Subscribers in total	Subscription fee
1	The United States	*New York Times*	3,300,000	$2/week
2	The United States	*Wall Street Journal*	1,500,000	$19.5/week
3	The United States	*Washington Post*	1,200,000	$1.25/4 weeks
4	The United Kingdom	*Financial Times*	740,000	$3.99/week
5	The United Kingdom	*The Guardian*	570,000	–
6	Japan	*Nihon Keizai Shimbun*	559,000	JPY¥4200/month
7	The United Kingdom	*The Economist*	430,000	£55/quarter
8	Germany	*Bild*	423,000	€7.99/month
9	The United Kingdom	*The Times*	260,000	£26/month
10	Sweden	*Aftonbladet*	250,000	SEK69/month
11	China	*Caixin*	200,000	$20.99/month
12	Brazil	*Folha de Sao Paulo*	192,000	R$19.90/month
13	France	*Le Monde*	180,000	€9.99/month
14	Poland	*Gazeta Wyborcza*	170,000	Zł 19.90/month
15	The United States	*The New Yorker*	167,000	$100/year
16	Sweden	*Dagens Nyheter*	150,000	SEK119/month
17	Norway	*Verdens Gang*	150,000	NOK189/quarter
18	France	*Mediapart*	140,000	€11/month
19	Italy	*Corriere della Sera*	133,000	€2.50/month
20	The United States	*Los Angeles Times*	133,000	$1.99/week
21	The United States	*National Geographic Magazine*	123,000	$12/year
22	France	*Le Figaro*	110,000	€9.9/month
23	Norway	*Aftenposten*	108,000	SEK249/month
24	Germany	*Die Zeit*	105,000	–
25	Argentina	*Clarin*	100,000	$55/month
26	The United States	*The Athletic*	100,000	$9.99/month
27	The United States	*Boston Globe*	100,000	$27.72/4 weeks

It is noticeable that since Internet media such as Facebook and Google can use news content from traditional media for free, advertisers are leaving newspapers for Internet media platforms, causing a drop in the profits among traditional journalism. It is expected that in the next few years, Internet media companies will balance their relationship with traditional news outlets by providing financial support and paying copyright fees.[6] According to *Wall Street Journal*, Facebook has proposed to ABC News, Bloomberg, Dow Jones, *Washington Post* and other giants in press and

[5] *2019 Global Digital Subscription Snapshot* released by FIPP and CeleraOne, visit https://d1ri6y1vinkzt0.cloudfront.net/media/documents/2019%20Global%20Digital%20Subscription%20Snapshot_1April.pdf.

[6] Digital News Report 2019, Reuters Institute, https://reutersinstitute.politics.ox.ac.uk/sites/default/files/2019-06/DNR_2019_FINAL_0.pdf.

Table 5.4 Summary of users getting news from different media[7]

Media type	User by percentage (%)	Female (%)	Male (%)
Online media (incl. social media)	82	83	81
TV (broadcast TV and cable TV)	70	70	70
Social media (incl. IM software)	52	55	50
Paper media	32	29	35
Radio station	32	29	35

Note The percentage is calculated based on the total number of Internet users

publishing that it is willing to pay $3 million in copyright fee per year for the use of content generated by traditional media in its upcoming news program.

5.2.2.2 Social Media News

Social media have become an important channel for acquiring news. Although TV is still the most widely used channel for news, there has been a steady increase in the number of people using Internet and social media to get news. Summary of users getting news from different media are listed in Table 5.4.

Pew Research Center has been tracking changes in preferred news channels for news among American adults and found that in 2018, the percentage of users obtaining news from social media surpassed that from newspapers for the first time, at 20% and 16%, respectively.[8] In the United Kingdom, TV remains the most commonly used channel to get news, followed by Internet, radio, and newspapers. And 49% of adults obtain news through social media.[9]

(1) More and more young users are using social media for news. According to Reuters Institute for the Study of Journalism, University of Oxford, a survey of 38 markets finds that that 66% of young people aged 18–24 get news from social media, while among those aged 55 and above the number is only 43%. In the United States and the United Kingdom, social media are the most popular

[7]Digital 2019: Q3 Global Digital Statshot, 2019.7.17, https://datareportal.com/reports/digital-2019-q3-global-digital-statshot.

[8]Social media outpaces print newspapers in the U.S. as a news source, 2018.12.10, https://www.pewresearch.org/fact-tank/2018/12/10/social-media-outpaces-print-newspapers-in-the-u-s-as-a-news-source/.

[9]News Consumption in the UK: 2019, 2019.7.24, https://www.ofcom.org.uk/__data/assets/pdf_file/0027/157914/uk-news-consumption-2019-report.pdf.

news platform[10] among young people (18–29 years old in the former and 16–24 years old in the latter). Facebook, YouTube, Instagram, WhatsApp and Snapchat are the top five platforms for news among people aged 18–24.

(2) Press release platforms diversify. A survey of 200 editors, CEOs and digital leaders around the world shows that Google is the most favored publishing platform among news production organizations. 87% of the respondents considered Google as "important" or "very important", while 43% considered Facebook as "important" or "very important". Apple News and YouTube were considered equally important. In order to attract new users, news release agencies also set their eyes on Instagram and Twitter.[11]

(3) More and more users are using instant messaging platforms for news, making privacy a new trend in news consumption. Compared with the same period in 2018, people in many countries are spending less time on Facebook but more on WhatsApp and Instagram. WhatsApp has become the main channel for discussing and sharing news in Brazil, Malaysia, South Africa and other countries. In Turkey and Brazil, more users prefer to discuss news and politics in public or private Facebook groups.

5.2.3 *Online Entertainment*

5.2.3.1 Video on Demand (VOD)

Video on demand (VOD) roughly falls into three categories: VOD by subscription, video downloading and pay-per-view VOD. Among them, VOD by subscription refers to VOD services based on subscription; video downloading refers to video content that can be bought for permanent use; pay-per-view VOD refers to VOD services provided per payment, that is, once the payment is made, one can visit the video within a certain time frame from different devices.

According to *Digital Media Report 2019: Video-on-Demand* released by Statista, a statistics website, global VOD market in 2018 was about $30.5 billion (in terms of revenue), accounting for 21.2% of the total digital media market, as the second largest sector behind the video game business. The United States, Europe, and China accounted for 75.1% of the global market. The United States ranked the first with $13.8 billion in revenue, followed by Europe ($6.8 billion) and China ($2.3 billion). Among the three types of VOD services, VOD by subscription performed well, with

[10]Integrated Data Source: the UK data: News Consumption in the UK: 2019, 2019.7.24, https://www.ofcom.org.uk/__data/assets/pdf_file/0027/157914/uk-news-consumption-2019-report.pdf. The US data: More people get their news from social media than newspapers, 2018.12.10, https://www.engadget.com/2018/12/10/more-people-get-news-from-social-media-than-newspapers/?guccounter=1.

[11]For 2019 news, media and technological trends and forecast reports, visit https://reutersinstitute.politics.ox.ac.uk/sites/default/files/2019-01/Newman_Predictions_2019_FINAL_2.pdf.

Table 5.5 Market share of the top five VOD providers in seven countries

US	China	Germany	UK	France	Spain	Italy
Netflix 75%	IQiyi 65%	Amazon 66%	Amazon 55%	Netflix 51%	Netflix 64%	Netflix 55%
Amazon 56%	Tencent Video 51%	Netflix 47%	Sky Go 23%	Amazon 24%	Amazon 40%	Amazon 45%
Hulu 35%	YouTube 34%	Sky Go 15%	iTunes 23%	StudioCanal 23%	HBO GO 30%	Google Play 22%
HBO GO 31%	Mango TV 22%	Maxdome 12%	Google Play 22%	Google Play 14%	Movistar+ 26%	Infinity 19%
Google Play 30%	BaoFeng Player 16%	Google Play 9%	Now TV 20%	Orange 13%	Google Play 21%	CHILI 14%

$23 billion in revenue in 2018, while the revenue of video downloading and pay-per-view VOD grew relatively slower. Statista predicts that from 2018 to 2023, the average annual growth rate (AAGR) of global VOD business will reach 4.1%. Among them, the AAGR in China will go above 4.7%, in the United States above 3.6%, and in Europe above 3.4%.

There are obvious agglomeration effects among global VOD suppliers. According to a survey conducted by Statista, the top five VOD suppliers in the United States are Netflix, Amazon, Hulu, HBO GO and Google Play; the top five suppliers in China are iQiyi, Tencent Video, Youku, Mango TV and Baofeng Player; the top five suppliers in Germany are Amazon, Netflix, Sky Go, Maxdome and Google Play; the top five suppliers in the United Kingdom are Amazon, Sky Go, iTunes, Google Play and Now TV; the top five suppliers in France are Netflix, Amazon, StudioCanal, Google Play and Orange; the top five suppliers in Spain are Netflix, Amazon, HBO (Spain), Movistar+ and Google Play; and the top five suppliers in Italy are Netflix, Amazon, Google Play, Infinity and CHILI.[12] Generally speaking, Amazon dominates the market in Germany and the United States; Netflix is uniquely-positioned in the United States and maintains high market share in Spain and Italy; video providers such as iQiyi and Tencent Video occupy the Chinese market. See Table 5.5 for market share of the top five VOD providers in seven countries.

5.2.3.2 Digital Music

In 2018, global digital music market was approximately $12.7 billion (in terms of revenue), accounting for 8.9% of the total digital media market.[13] Currently, digital music is mainly divided into two categories: music streaming (online music)

[12]Digital Media Report 2019—Video-on-Demand, Statista, June 2019. The survey collects data by counting "In the past 12 months, which online video-on-demand providers have you used and paid for?"

[13]Data Source: Digital Media Report 2019—Digital Music, Statista, Released: June 2019, 见https://www.statista.com/study/39314/digital-music-2018/.

Table 5.6 Market share of the top five digital music providers in seven countries[14]

US	China	Germany	UK	France	Spain	Italy
Amazon 49%	Tencent Music 47%	Amazon 53%	Amazon 40%	Deezer 45%	Spotify 60%	Spotify 44%
iTunes 43%	KuGou 43%	Spotify 34%	Spotify 38%	Spotify 29%	Amazon 34%	Amazon 30%
Spotify 41%	NetEase 35%	iTunes 18%	iTunes 37%	Amazon 23%	Google Play 27%	iTunes 28%
iTunes 36%	Baidu 33%	iTunes 14%	iTunes 27%	iTunes 23%	iTunes 23%	Google Play 23%
Google Play 29%	CoolWo 17%	Google Play 9%	Google Play 17%	Napster 8%	iTunes 22%	iTunes 22%

and audio downloading. There are two profit models for music streaming, the subscription-based profit model (such as Spotify's value-added services and iTunes's subscription service, no advertisements), and advertisement-based profit model (such as Spotify's free service). Audio downloading is a digital music service that provides paid downloads of singles, albums or arrangements. According to *Global Music Report 2019* released by International Federation of the Phonographic Industry (IFPI), by the end of 2018, the number of users of paid music streaming services worldwide reached 255 million, representing a 34.0% growth year on year in revenue, which contributed to 47% of the whole music industry. Audio downloading suffered a 21.2% drop in revenue year on year, with continued decline in market share.

According to the statistics of Statista, the digital music market in the United States, Europe, and China totaled $9.4 billion in 2018, accounting for 73.6% of the global digital music market. Among them, the United States contributed $5.2 billion, Europe $3.4 billion, and China $0.8 billion. Digital music business in Germany reached $760 million in revenue, the highest among the five major European markets.

In the music streaming market, Amazon Music ranks the first in the United States, Germany and the United Kingdom, followed by Apple Music and Spotify, and Tencent Music is the most popular one in China.

See Table 5.6 for market share of the top five digital music providers in seven countries.

5.2.3.3 Video Games

In 2018, global video game market was approximately $79.2 billion (in terms of revenue), accounting for 55.2% of the total global digital media market. The video game market in China, the United States and Europe totaled $52.2 billion, accounting

[14]Calculated as the percentage of users purchasing digital music from online suppliers in the past 12 months.

for 67.5% of the whole world. Among them, China ranked the first with $22.1 billion in revenue, and the United States came second with $17.5 billion, followed by Europe with its combined revenue of $12.6 billion. Among all kinds of game products, mobile games (including mobile phones and tablets) generated $51.1 billion, constituting the highest contribution to the market revenue.[15]

In the field of mobile games, based on the number of monthly active users in the second quarter of 2019, four games from China made into the top five list, and two from the United States and two from Finland ranked in the top ten. Over the same period, Stack Ball, Run Race 3D, Free Fire, Subway Surfers and Color Bump 3D were the top five downloads in the main application market. See Table 5.7 for Ranking of Mobile Games in the Second Quarter of 2019.

Table 5.7 Ranking of mobile games in the second quarter of 2019[16]

Rank	By monthly active user		By download	
	Game	Developer	Game	Developer
1	PUBG Mobile	Tencent (China)	Stack Ball	AZUR (Russia)
2	Candy Crush Saga	Activision (The United States)	Run Race 3D	Good Job Games (Turkey)
3	Honour of Kings	Tencent (China)	Free Fire	Garena (Singapore)
4	Game for Peace	Tencent (China)	Subway Surfers	Kiloo Games (Denmark)
5	ANIPOP	Happy Elements (China)	Color Bump 3D	Good Job Games (Turkey)
6	Pokémon Go	Niantic, Inc. (The United States)	Tiles Hop: EDM Rush	AMANOTES (Vietnam)
7	Clash of Clans	Supercell (Finland)	Clean Road	SAYGAMES (Belarus)
8	Clash Royale	Supercell (Finland)	PUBG Mobile	Tencent (China)
9	Subway Surfers	Kiloo Games (Denmark)	Traffic Run	Geisha Tokyo (Japan)
10	Helix Jump	VOODOO (France)	Crowd City	VOODOO (France)

[15]Data Source: Digital Media Report 2019—Video Games, Statista, Released: Aug 2019, https://www.statista.com/study/39310/video-games-2018/.

[16]Source: Hootsuite Digital 2019, Q2 Global Digital Statshot.

5.3 Development and Application of Internet Media

Technological progress has profoundly changed the production method, communication pattern and ecosystem of Internet media. The application of 5G, cloud computing, artificial intelligence and other technologies in Internet media has boosted information use and sharing rate and content production efficiency, improved communication effect and user experience, and optimized the media value chain.

5.3.1 5G Improving Media Production and Communication

2019 is the first year of commercial uses of 5G. It is estimated that by the end of 2019 there will be over 10 million 5G network users worldwide.[17] 5G technology can effectively boost news production and dissemination, improve user experience with news consumption, and further inspire new business forms and applications.

(1) At news gathering and editing stage, 5G technology can generate new ideas and expand the space for news production. Featured by high speed, high reliability, low latency and large capacity, 5G technology effectively improves the speed and quality of real-time collection and transmission of massive information and data, which will enhance reporter's ability to capture and produce news and change news production form deeply. First, a novel news Rewriting model[18] is emerging. Reporters can use 5G high-definition video to stream data back to the news editing department in real time, generate new cases based on big databases, and then edit raw materials into news through machine learning, computer vision, artificial intelligence and other technologies. Second, 5G networks and artificial intelligence make it easier to actualize sensor journalism. Sensor journalism is a news production model that relies on sensors for data collection and uses data processing technology as support. During early testing, sensor journalism faced obstacles such as the high network access cost involving many small devices and the lack of raw data analysis tools. The application of 5G networks and artificial intelligence will strengthen data transmission and analysis capabilities,[19] making it cheaper and faster with higher analytical accuracy.

---

[17]Canalys: 1.9 billion 5G smartphones will ship in the next five years, overtaking 4G in 2023, https://canalys.com/newsroom/5G-forecasts-five-year#.

[18]The long-standing tradition of rewriting refers to the method that a group of reporters working on one report send their findings to person in the office, who then assembles and organizes the raw materials into the final coherent report.

[19]Data Source: JOSHUA BENTON (2019) What will journalism do with 5G's speed and capacity? Here are some ideas, from The New York Times and elsewhere, visit https://www.niemanlab.org/2019/04/what-will-journalism-do-with-5gs-speed-and-apacity-here-are-some-ideas-from-the-new-york-times-and-elsewhere/.

(2) At news dissemination stage, 5G technology can provide users with more immersive experiences. The combination of 5G and AR/VR technology provides viewers with no-delay, immersive, high-definition virtual experience, and is more widely used in live TV streaming. For example, in 2019, China Central Television used "5G + 4K" ultra-high-definition video real-time back live streaming model and "5G + VR" immersive virtual technology to broadcast Spring Festival Gala live. South Korea's SKT applies 5G to the baseball stadium for VR live broadcast. American T-Mobile explores AR remote operation and control technology in 5G settings. The United Kingdom promotes 5G firstly in VR live broadcast and provides VR boxes to audiences who are present at sports events like Premier League and to those who watch games on fixed broadband TV at home, giving them multi-angle viewer experience.[20]

5.3.2 Cloud Computing Changing Media Production Ecology

In media production ecosystem, cloud computing is mainly used to integrate content channels, share resources and improve efficiency, which helps virtualize content production and bears such advantages as flexibility and scalability, immersive and dynamic content production and distribution, and accessibility anytime and anywhere. For example, "China Blue Cloud" Platform developed by International Film and Television Center of Zhejiang Radio and Television Group is a multi-tenant-based hybrid media cloud that provides integrated services including "collection, editing, dissemination, management, storage and use" in media production, and contributes to an efficient and reliable media production ecology. Hubei Radio and Television Station in China establishes the Yangtze River Cloud Platform, which consolidates media terminals and products at county, city and province levels through its "Cloud Report Library" and the provincial "Central Kitchen" and builds a regional ecological media convergence platform featured by "news + government + services".

Cloud computing facilitates real-time sharing and processing of information. Upon collecting video materials for breaking news, news team on the scene can immediately send them back to the news editing department through cloud-based content sharing platform. For example, Avid's new-generation media workflow platform "MediaCentral | Cloud UX" provides full cloud or local/cloud hybrid storage solutions mainly for TV news, sports events and post-production business. It provides optimization and design services to reporters for the timeliness of news, and helps them efficiently collect and edit news and deliver it on TV or social platforms at the earliest possible time.[21]

[20]Prospects of global trends in 5G applications: joint rise of applications and traffic flow, visit https://stock.jrj.com.cn/invest/2019/08/07074527940565.shtml.

[21]Data Source: Avid MediaCentral | Cloud UX, visit https://avid.force.com/pkb/articles/en_US/readme/MediaCentral-Cloud-UX-v2019-Documentation.

At present, IT giants such as Amazon, IBM, Google and Microsoft compete fiercely in the field of cloud computing and its virtualization functions. They not only attach great importance to the development of network applications and platforms with media attributes, but also apply cloud computing to media services through cooperation with other companies. For example, in 2019, Microsoft's Azure and Rohde and Schwarz worked together and launched Prismon Cloud with built-in Azure services. It was a cloud computing-based Over the Top (OTT) media service monitoring solution that provided signal analysis, monitoring and quality control of audio and video content to ensure excellent media experience for users.[22]

5.3.3 Artificial Intelligence Optimizing Media Value Chain

Artificial intelligence will affect all aspects of media value chain such as supply, output, and consumption. It will boost content production efficiency, assist media companies in decision-making, and improve the effectiveness of information dissemination.

5.3.3.1 Using AI to Deal with Information Overload

Artificial intelligence can help reporters catch news clues, verify materials, and save manpower, time and cost. Reuters has developed Lynx Insight, an automated reporting tool, and News Tracer, a news tracking tool, for reporters to mine report topics, filter clues and drop unreliable sources, thereby improving news production and reporting efficiency. In November 2018, Quartz, a business news website, launched Quartz AI Studio, an open source platform. The machine learning tools provided by the platform can help reporters analyze data,[23] freeing them from massive data processing and letting them focus on the production of effective news.

5.3.3.2 AI Inspiring Innovative on Form of Content Products

Apart from automatic text generation robots, progress with AI virtual anchor has been made. In 2018, China's Xinhua News Agency and Sogou jointly launched the world's first AI-synthesized News Anchor based on AI simulation technology. In 2019, The Paper and Baidu worked together to develop AI-powered virtual anchors and

---

[22]Data Source: R&S®PRISMON Audio/Video Content Monitoring and Multiviewer Solution, https://cdn.rohde-schwarz.com.cn/pws/dl_downloads/dl_common_library/dl_brochures_and_datasheets/pdf_1/PRISMON_bro_ en_5214-8454-12_v1100.pdf.

[23]Data Source: Quartz AI Studio launches an open-source platform to help journalists use machine learning, https://www.journalism.co.uk/news/quartz-ai-studio-launches-an-open-source-platform-to-help-journalists-use-machine-learning/s2/a732936/?fbclid=IwAR29TwJmT1cGs03v9gXlZpK1u1SLFXVGcuxAHMkY50l9FS3JbLfWB6CVzIk.

produced their first daily news program broadcast by real-life-like virtual anchors.[24] Japan's national public broadcasting company NHK created Yomiko, an animated news anchor to broadcast the evening news.

5.3.3.3 AI-Aided Internet Platform Content Review

AI integrates deep learning, computer vision algorithms, natural language processing, speech recognition and other technologies to analyze different types of information such as text, voice and video on Internet platforms. Based on massive sample training, it can identify pornography, violence, terrorism and other ill-intended information on the platform, and reduce the labor cost of content review to a certain extent.

Notably, though AI is of great help to news production and customized recommendation, it has also created ethical, moral and legal challenges. At present, most concern is about how to make sure news media use AI technology responsibly and transparently and prevent them from reader manipulation. In April 2019, European Commission issued *AI Ethics Guidelines*.[25] In June 2019, China's Professional Committee on New-Generation AI Governance issued *Principles of Next-Generation AI Governance*,[26] which proposed guidelines to ethically regulate AI technology.

5.4 Governance and Challenges of Internet Media Content Ecosystem

Though the rapid development of new technologies and applications in media vitalizes media content, it also makes content ecosystem complicated and adds challenges to the governance of media ecosystem.

5.4.1 New Changes in Online Dissemination of False Information

Social media, especially personal instant messaging services, are turning into a main channel for the spread of false information on Internet. As network technology advances, there are more channels for spreading false information on Internet, especially in the age of social media. Since netizens are both information receivers and

[24] "A Virtual Anchor Co-Developed by Baidu Brain and The Paper up and Running", https://www.chinanews.com/business/2019/07-22/8903273.shtml.

[25] *AI Ethics Guidelines* released by EU, visit https://www.xinhuanet.com//tech/2019-04/15/c_1124365850.htm.

[26] *Principles of Next-Generation AI Governance* released by China, visit https://www.xinhuanet.com/tech/2019-06/18/c_1124636003.htm.

disseminators, their large number has facilitated the dissemination of false information, which is cheap, fast and with extensive impact. Naturally, social media are an ideal "hotbed" for false information. In March 2019, a survey in the United States showed that nearly two thirds of the respondents believed that the spread of "false information" and "misinformation" was one of the main problems faced by the United States, and 55% of the respondents considered fake social media accounts were the main channel for dissemination of false information. Among them, 64% of respondents believed that Facebook was responsible for dissemination of false information, followed by Twitter (55%), YouTube (48%), Instagram (46%), Snapchat (39%) and LinkedIn (28%).[27]

Rooted in their social networking function, personal instant messaging software is growing more powerful at spreading news and other information, and inevitably more frequently used to disseminate false information. A survey conducted by Reuters Institute for the Study of Journalism in June 2019 shows that WhatsApp is becoming a major tool for news and information dissemination in some countries. For example, in Brazil, 53% of the users check it for news and information, followed by 50% in Malaysia and 49% in South Africa. A study on India and Brazil finds that WhatsApp is involved in the rapid spread of misinformation, political propaganda and hatred speech. Researchers analyze 100,000 images circulated among 347 WhatsApp groups during the 2018 Brazil presidential election and find that only 8% of the images are completely real.[28] A survey conducted by Reuters Institute for the Study of Journalism on how digital news is used in India finds that 57% of respondents question the authenticity of online news, more than half of the respondents express concerns about false information and fake news, and WhatsApp is considered to be one of the main channels for spreading rumors.

Deep fake has caused new challenges to the control of false information on Internet. Built on AI's strength in deep learning, language recognition, image recognition, big data processing and other capabilities, deep fake technology can imitate human physiological characteristics and synthesize false audio and video, which produces highly convincing imitation indistinguishable from the real ones and further promotes the flooding of false information. In January 2019, a fake video drew people's attention to deep fake technology. The video that showed the former US President Barack Obama was giving a speech was actually forged by synthesizing a speech given by an actor imitating Obama's voice with the image. False information created by deep fake technology poses serious potential threats to political security, public security, and national security. Governments and Internet companies are studying and taking measures to deal with the according challenges. U.S. Congress has held multiple hearings on deep fake and online misinformation. Congressmen from both parties promoted *Deepfakes Report Act of 2019*, demanding Department

[27]Data Source: *2019 IPR Disinformation in Society Report*. The Institute for Public Relation, visit https://instituteforpr.org/ipr-disinformation-study/.

[28]Data Source: the WhatsApp election. Released by India, Financial Times, visit https://www.ft.com/content/9fe88fba-6c0d-11e9-a9a5-351eeaef6d84.

of Homeland Security to issue a report on deep fake and relevant content every year.[29] Facebook, Google, Twitter and other Internet companies have also stated that they are studying and formulating countermeasures specifically for deep fake technology.

Where there lacks accurate and authoritative information, false and misinterpreted information will cause public disruption. Faced with new changes and features of online misinformation, many governments have introduced countermeasures, especially "Action Plan against Disinformation" initiated by European Commission in December 2018. The plan aims to safeguard the 2019 European Parliament Election and elections in its member states or regions in 2020 against misinformation, enhance EU and its members' ability to protect themselves against false information, and maintain political stability and defend national security within EU.[30]

5.4.2 Hard to Control the Spread of Terrorist and Extremist Content Online

Terrorist and extremist content is hard to control and is subject to various limitations such as technical obstacles, legal restrictions and public concerns. Facebook pointed out that it was difficult to distinguish hatred speeches from legal political ones. Twitter once closed the accounts of nationalists, but was pressured to reopen them later. YouTube was accused of incompetence for handling videos that propagated Neo-Nazi's "National Action". In 2019, many incidents where Internet media were used to spread cyber terrorism and extremism drew attention around the world. In March 2019, an assailant broadcast the whole assault of a shooting in a mosque live on Facebook. The pictures (including gifs) and even uncensored video, after being posted on social media, ended up on the front page of news websites with global influence and were forwarded by many viewers. Although the original video was quickly deleted, its copies had already been uploaded onto other platforms such as YouTube and Twitter, making it very difficult to stop the spread. The shooting highlighted the necessity and urgency of the combat against abusing Internet to spread hatred and incite violence. Many people believed that social media should be more serious about and work harder to deal with the threat of cyber terrorism and extremism.[31]

The shooting directly led to the signing of a non-binding agreement "Christchurch Call to Action" among Canada, EU, New Zealand, Senegal, Indonesia, Jordan and other countries and regions. And Facebook, Google, Twitter and other Internet

---

[29]Data Source: *Deepfakes Report Act of 2019* released by H.R.3600, Visit https://www.congress.gov/bill/116th-congress/house-bill/3600/text?q=%7B%22search%22%3A%5B%22deepfakes%22%5D%7D&r=1&s=1.

[30]Data Source: European Commission (2019) Action Plan against Disinformation. Brussels, 5.12.2018 JOIN (2018) 36.

[31]Data Source: Christchurch shootings: Social media races to stop attack footage, BBC News, visit https://www.bbc.com/news/technology-47583393.

companies promised to develop a series of tools to prevent the upload of terrorist and extremist content on the platform, improve the transparency of content deletion and detection, review and modify social media algorithms and data sharing, and fight the root of violence and extremism, to stop the posting of terrorist and extremist content online and prevent Internet media from being abused by terrorists.[32] In April 2019, EU passed *Regulation on Preventing the Dissemination of Terrorist Content Online*, stipulating that hosting service providers should delete or block the access to certain content within one hour after receiving such instructions from authorities of member states. Australia passed *Sharing of Abhorrent Violent Material Act* to regulate social media. The British government funded and established Social Media Center in response to the abuse of social media for promoting gang culture and inciting violence by criminal organizations. Japan strengthened the supervision of harmful information such as online suicide. The U.S. Department of Homeland Security worked with major social media companies and encouraged them to enforce corporate supervision of online content.[33]

So far, major social media companies have stepped up their countermeasures against online terrorist and extremist content. Facebook will shut down links to illegal content and other websites to prevent the spread of inappropriate content. YouTube focuses on technology's vital role in fighting extremist content online and cuts down the recommendation of extremist content by up to 50% by changing recommendation algorithms.[34]

5.4.3 Drawback of Media Platform Monopoly

In the boom of Internet economy, numerous large Internet platforms emerged. Armed with huge user markets, a wide range of products and services, and leading technologies and business models, they have exerted huge influence and even acquired dominance in some specific fields. In Internet media industry, in particular, companies such as Google and Facebook have built "entry barriers" against other competitors with their monopolistic advantage, which severely inhibits industrial innovation and arouses risks like privacy leakage. In 2019, regulatory authorities of many countries launched anti-monopoly investigations and imposed penalties on large Internet media platforms with different focuses.

---

[32] Data Source: A Global Call to End Online Extremism, visit https://foreignpolicy.com/2019/05/15/a-global-call-to-end-online-extremism-christchurch-jacinda-ardern-macron-violent-extremism-twitter-google-facebook-sri-lanka/.

[33] "We do see some results from these efforts. I believe we still have a long way to go, and we look forward to continuing our work with social media in this environment." Brian Murphy–Principal Deputy Undersecretary, Department of Homeland Security, Testimony to USA Congress, 5/19.

[34] Data Source: Susan Wojcicki, Interview @ Code Conference, visit https://www.vox.com/recode/2019/6/11/18660779/youtube-ceo-susan-wojcicki-code-conference-peter-kafka-interview-transcript-maza-crowder-lgbtq.

The United States takes cautious and tolerant attitude in regulating large Internet platforms. Both Federal Trade Commission (FTC) and Department of Justice place more emphasis on encouraging platform innovation in terms of product services, technological R&D, and business models, which expands traditional regulatory goals that limited to protecting consumers and maintaining market competition, and creates room for better platform development.[35] However, since the early 2019, the United States has tightened its regulatory measures on Internet platforms. The Subcommittee on Antitrust and Monopoly of Judiciary Committee of the House of Representatives held hearings on suspected monopoly of several large Internet companies. Department of Justice launched antitrust investigations into American tech giants such as Google and Facebook. Yet opinions differed on the implementation of specific antitrust measures. Makan Delrahim, an official from Department of Justice, believed that to judge whether a company break antitrust laws, the investigation must focus on whether there was any increase in price, drop in quality, damage to privacy, or exclusive agreements designed to hurt competitors.[36] Tim Wu, a legal expert at Columbia University, believed that corporate break-up was a way to end monopoly of Internet giants.[37] David Ciciline, the House Antitrust Subcommittee Chairman, believed that corporate break-up was the last resort, and that legislative or regulatory reforms were the key.[38]

EU has imposed very strict anti-monopoly regulations on large Internet platforms, and focused on the protection of the interests of other competitors and small-and-medium-sized firms in the market, with fines and taxation as its commonly used methods. From 2017 to 2019, Google was punished by EU for three consecutive years, totaling over 8 billion euros in fine for its dominance in the search engine market. In April 2019, the European Parliament decided that should social network companies such as Facebook and Twitter fail to remove extremist content quickly after receiving such instructions from regulatory authorities, they could be subject to a fine of up to 4% of their global operating revenue. In July 2019, German regulatory authorities concluded that Facebook underreported the amount of illegal content on the platform, which was a violation against their regulations on Internet transparency, and therefore imposed a €2 million fine on Facebook. In August 2019, EU launched an investigation[39] into the "potential anti-competitive behavior" of Libra, a digital currency developed by Facebook.

---

[35]Hongru [1].

[36]Senior US Antitrust Official: Splitting Google and Facebook, Taking a Lesson from History Sina Tech, visit https://tech.sina.com.cn/i/2019-06-12/doc-ihvhiqay5054037.shtml, June 12, 2019.

[37]Data Source: Tim Wu Explains Why He Thinks Facebook Should Be Broken Up, visit https://www.wired.com/story/tim-wu-explains-why-facebook-broken-up/.

[38]The U.S. Department of Justice Opened an Investigation into Internet Companies Suspected of Monopoly, Four Giants Refused to Comment. visit https://www.xinhuanet.com/world/2019-07/25/c_1210212601.htm, July 25, 2019.

[39]Data Source: Facebook's Libra Currency Gets European Union Antitrust Scrutiny, visit https://www.bloomberg.com/news/articles/2019-08-20/facebook-s-libra-currency-gets-european-union-antitrust-scrutiny.

The United Kingdom launched a special investigation into large Internet platforms to determine whether they had set up industrial competition barriers. In July 2019, the UK Competition and Markets Authority (CMA) stated that it was assessing whether online platforms were harming consumers and competitors in the digital advertising market. The investigation focused on whether consumers could fully control personal data as well as the data liquidity methods on Internet platforms. The survey specifically mentioned Google and Facebook, saying that these two companies accounted for 61% of the country's digital advertising revenue. CMA would determine whether Google and Facebook's dominance in digital advertising restricted the entry and competition of other providers. The results of the investigation will be announced in July 2020 at the latest.

Australia established a special agency responsible for platform anti-monopoly supervision investigations. In December 2018, Australian Competition and Consumer Commission (ACCC) released *Digital Platforms Inquiry: Preliminary Report*, pointing out that Facebook and Instagram accounted for 46% of Australia's advertising market revenue on digital display, with an average of 17 million users surfing on Facebook and 11 million users on Instagram every month. In response to the unfair competitive advantage caused by platform monopoly, the report put forward 23 recommendations, including clarifying the terms of unfair contracts, updating and improving merger laws and procedures, giving consumers more choice in browsers and search engines, imposing supervision on advertising and related businesses as well as news and digital platforms, and conducting review on the media regulatory framework.[40] In July 2019, Australian antitrust regulatory authority announced that it would set up a special agency within ACCC to supervise large Internet platform companies and investigate how they used massive data resources and advantageous algorithms on the platform to deliver advertisements to users and earn huge revenues.

In addition, antitrust agencies in Argentina, Canada, Brazil, Israel, China's Taiwan, South Korea, and Russia have also filed antitrust lawsuits against Google. Large-scale Internet platform companies, for their large user groups and massive data resources, draw more and more attention from national authorities on monopoly issues and probably face increasingly stricter regulations on a global scale.

Reference

1. Xiong Hongru, "Main Challenges and International Practice of Anti-Monopoly Regulations in the Digital Economy Era", *Economic Review Journal*, Vol. 07, 2019.

[40]Data Source: Digital Platforms Inquiry: Preliminary report, visit https://www.accc.gov.au/system/files/ACCC%20Digital%20Platforms%20Inquiry%20-%20Preliminary%20Report.pdf.

Chapter 6
Development of World Cybersecurity

6.1 Outline

The world today is featured by major development, major change and major restructuring with growing instability and uncertainties. As an important part of non-traditional security, cybersecurity becomes increasingly relevant to the common interests of mankind, world peace and development, and the national security for countries around the globe. Looking back on global cybersecurity in 2019, traditional and emerging threats intertwined, domestic and international communities highly interrelated, and online and offline environments interacted closely, with threats and risks on cybersecurity growing more prominent. Countries around the world attach more attention to cybersecurity and continuously improve capability and level on cybersecurity protection in terms of strategic planning, top-level design, institutional systems, technological capability and international cooperation.

1. High Prevalence of Cybersecurity Threats

As one of the greatest cybersecurity threats with global impact in the past two years, ransomware has attacked high-value targets in many countries. Drawn by the rebound in Bitcoin price, Mining Trojan becomes one of the most widespread malwares. Under the frequent occurrence and high incidence of Advanced Persistent Threat (APT) attacks, the supply chain falls into new favored target. The security of critical information infrastructure faces growing pressure, and key industries such as electric power and industrial Internet suffer from more cyberattacks. Large-scale data security incidents become more common due to poor awareness of cybersecurity protection, insufficient capability, and improper operations. The development of new technologies such as artificial intelligence, Internet of Things, and blockchain broadens and blurs the boundary of cybersecurity, and the potential risks therein deserve great attention.

2. More Effort into Building Cybersecurity Protection Capability

Countries around the world have formulated and improved cybersecurity strategic plans, laws and regulations, stepped up cybersecurity management systems and

Chinese Academy of Cyberspace Studies, *World Internet Development Report 2019*, https://doi.org/10.1007/978-981-33-6938-2_6

mechanisms, strengthened security protection of critical information infrastructure, enhanced the protection of data security and personal information, developed new cybersecurity protection technologies, accelerated the cultivation of cybersecurity talent, and deepened international cybersecurity cooperation, which further boosts the overall cybersecurity protection level.

3. Rapid Development of Cybersecurity Industry

Countries around the world have prioritized the development of cybersecurity industry, optimized policies and measures and increased investment to accelerate the development of cybersecurity industry, promote continuous expansion of relevant enterprises, and maintain sustainable growth of the field.

4. Increasingly Fierce Arms Race in Cyberspace

Countries around the world take cyberspace as a national strategic priority and a high ground worth competing for. They quicken the strategic deployment of cybersecurity, enhance the construction of cyber defense, and strengthen the capability in cyberwar. Certain countries heighten cyber deterrence strategy, which intensifies the arms race in cyberspace. Cyber warfare moves from theoretical hypothesis into real threat, and uncertainty and confrontation in cyberspace increase. Faced with the increasingly complex situation of international cybersecurity, it is urgent for all countries to deepen strategic mutual trust, strengthen communication and cooperation, and jointly maintain peace and security of cyberspace.

6.2 Major Threats to Cybersecurity in the World Today

Security threats that global cyberspace faces now display a multi-level, multi-dimensional and cross-domain trend. Traditional cybersecurity threats continue to evolve, data security risks increase, critical information infrastructure suffers organized and high-intensity cyberattacks, and new cybersecurity threats keep emerging, which pose great risks and challenges to national politics, economy, culture, society, defense, and the rights and interests of citizens in cyberspace.

6.2.1 Spread of Ransomware Threatening the World

As one of the greatest cybersecurity threats with global impact in the past two years, ransomware has attacked many countries in 2019. In terms of attack pattern, the choice of targets shifts from indiscriminate targets to designated high-value targets. In terms of attack feature, ransomware iterates quickly and spreads widely with diverse variants and high concealment, making it very difficult to track and prevent. In 2018, the most widespread ransomware GandCrab had at least 5 variants in just over a year, and it required the use of new cryptocurrency Dash for payment. This enlarged its concealment. In addition, GandCrab proliferated by the "ransomware as a service" model and infected more targets. Sodinokibi, a ransomware that broke out in April 2019, attacked hosting service providers through vulnerability distribution, abused system vulnerabilities for greater Windows access, and used legitimate processor functions to evade security solutions, which was rarely seen with previous ransomware. In terms of attack target, ransomware attacks key information infrastructure of important industries such as transportation, energy and healthcare, to disturb the normal operation of society. In mid-June 2019, ASCO, a global aircraft parts supplier, was attacked by a ransomware, which paralyzed its production system and put nearly 1000 employees on furlough. In terms of ransom, the amount is huge and growing. The operating team behind GandCrab claimed that in just one and a half years, they made $2 billion in profit.

6.2.2 *Mining Trojan Active Again with Widespread Threat*

Mining Trojan samples account for a large proportion of all Trojan viruses and are the most common viruses in recent years. In 2019, the price of Bitcoin soared, leading to the rebound of entire digital encryption currency industry. And mining Trojan, closely related to the currency market, entered a new active cycle and became one of the most widespread malwares. Surveys show that mining attack quadrupled compared with in 2018, covering almost all platforms. In their most active phase, mining Trojan generated up to 150,000 samples[1] on average every day. Notably, there is a growing trend of organized and industrialized operation of mining Trojan. In order to maintain long-term operation, mining Trojan is equipped with increasingly sophisticated function design. By combining themselves with botnets and ransomware, mining Trojan can achieve continuous iteration and fast updating. In addition, mining Trojan makes breakthroughs in cross-platform attack capability. In June 2019, researchers from the cybersecurity company ESET found that LoudMiner, a new type of malicious cryptocurrency mining Trojan, operated across macOS and Windows systems for continuous mining of cryptocurrencies through virtualization software. In the future, as mining Trojan continues to evolve in terms of hidden tactic, attack method and countermeasure capability, more disruptive variants will appear.

6.2.3 *High Incidence of APT Attacks*

In recent years, APT attacks are highly active, causing widespread concern.

(1) Asia and other regions are under serious APT attacks. According to *2018 APT Summary Report* released by 360 Threat Intelligence Center in January

[1] *The 2018–2019 Cybersecurity Observation Report* released by Venustech and Freebuf.

2019, South Korea, the Middle East, Pakistan, Japan, Ukraine, and China rank relatively high[2] among the regions most severely impacted by APT.

(2) There is a new trend of APT attacks on supply chains. Attacks on supply chains are easy to overlook among cyberattacks, but in recent years, there have been cases targeting game industry and network hard disk services.[3,4] The organizations that carry out APT attacks on supply chains have special intentions, targeting certain individuals or institutions as "curve attack", namely using the attack on relevant suppliers or service providers to get to their ultimate target.

(3) The leakage and proliferation of APT Cyber arsenal has raised concern. In March 2019, a member disclosed cyber weapons and related information of APT 34, a hacker organization, on Telegram. Later, some other hackers disclosed relevant information of Muddy Water through the same channel and put them up for public auction. The proliferation and abuse of cyberattack weapons have further systematized the attack capability of APT organizations and brought great risks to cyberspace security.

6.2.4 Frequent Attacks on Critical Information Infrastructure

6.2.4.1 Power Grid Infrastructure Becoming an Important Target

With network connection of power control systems deepening and IoT products like smart devices and sensors surging, plenty of low-security and high-risk infrastructure are completely exposed to Internet, which increases the risk of cyberattacks. In March 2019, a blackout occurred in Venezuela, which affected its capital Caracas and at least 20 of the country's 23 states. The incident was recorded as the longest power outage of the country since 2012 with the widest impact. The Venezuelan government claimed that the large-scale blackout was the result of a foreign cyberattack. In July 2019, a large-scale power outage in Manhattan, New York City of the United States, affected tens of thousands of residents, as well as businesses and some transportation facilities. The United States claimed that information warfare forces of other countries penetrated and attacked information station of the control center of more than 30 transformer substations in New York City.

---

[2] *2018 Summary Report of Global Advanced Persistent Threat (APT)* released by 360 Company, January 2019.

[3] Data Source: Marc-Etienne M. Léveillé, Gaming industry still in the scope of attackers in Asia [OL], 2019-3-11.

[4] Data Source: Anton Cherepanov, Plead malware distributed via MitM attacks at router level, misusing. ASUSWebStorage [OL], 2019-5-14.

6.2.4.2 Grim Outlook of Industrial Internet Security

Industrial Internet links the industrial system with Internet and expands a country's key information infrastructure from breadth to depth in a three-dimensional grid. As a result, cybersecurity and industrial security risks are intertwined and amplified. According to *Introduction to Industrial Information Security* released in May 2018, there are nearly 300 industrial information security incidents worldwide each year since 2015, and the industrial field is an emerging hot target for cyberattacks. Ponemon Institute, an American Security Research Center, issued a report in April 2019, stating that among the 701 critical infrastructure providers in public utilities, industrial manufacturing and transportation industries it surveyed, about 90% of the respondents had been exposed to cyberattacks, resulting in data leakage or facility shutdowns in the past two years.

6.2.5 Growing Data Security Risks Raising Concerns

The continuous development of technologies such as big data, blockchain, and Internet of Things has pushed the boundaries of the data life cycle, generated new technological architectures and tools for data processing, and deepened data integration through chat robots, facial recognition technology, algorithmic decision-making and other applications, which poses new challenges to the protection of data security.

(1) Data leakage is more frequent and on increasingly larger scales. According to *Global Risk Report* released by the World Economic Forum in Davos in 2019, data leakage ranked as one of the top five risks in the world in 2019 with an 82% chance of occurrence. According to Identity Theft Resource Center (ITRC), more than 1100 data leakage cases occurred in 2018, exposing over 56 million records in total. According to a data leakage report for the second quarter of 2019 released by a Russian security company InfoWatch, compared with the same period in 2018, the number of confidential corporate data leaked worldwide increased by nearly 28%. In May 2019, First American, an American mortgage insurance company, was report to have leaked information on nearly 885 million customers, including massive personal information, bank account numbers, social security numbers, driver license information, mortgage payments, and tax records, etc.

(2) Poor awareness of cybersecurity protection, incompetence, and misoperation has exacerbated the occurrence of data leakage. The data leakage of an American data company Exactis in June 2018 was not the work of some hacker attacking its database or other complex malware, but due to the lack of encrypted firewall on its servers, which exposed them directly to the public database. In August 2019, it was reported that the biometric security company Suprema Biostar suffered a 23 GB data leakage from its fingerprint recognition database. Security investigators found that the database lacked duly protection

and most of the data including the admin account password were stored without encryption.

(3) Concerns about data security grow around the globe. According to the survey results of *2019 Unisys Security Index* released by the American consultancy Unisys, global concerns on data security reached 175 points on the index in 2019, a record value in the past ten years, registering an increase of nearly 50% from 2009.

6.2.6 Emerging Security Risks of New Technologies and Applications

6.2.6.1 Prominent Cybersecurity Risks with Internet of Things

The cybersecurity protection for Internet of Things remains in its infancy, lacking in capability and staying underdeveloped in practice. IoT system faces threats from malware, Trojan viruses and malicious scripts.

(1) IoT devices face serious vulnerability threats. In order to control costs, many IoT device manufacturers deliberately ignore security concerns and fail to fix known vulnerabilities in a timely manner, leaving their products an easy target for cyberattacks. According to Fortinet's threat assessment report in the fourth quarter of 2018, half of the world's top twelve vulnerabilities are found in IoT devices.

(2) The integration of IoT, 5G, and cloud computing causes new problems to cybersecurity. 5G constructs a scenario where everything is connected, which also expands the range for cyberattacks and creates more opportunities for hackers.[5] In the future, there will be tens of billions of networked devices and each may serve an entry point for attacks, making cyberattacks extremely hard to prevent.

(3) Security issues of Internet of Things pose a serious threat to privacy protection. According to relevant data, each household can generate up to 15,000 discrete data points[6] per day, but smart home devices connected to Internet of Things generally do not have security protection functions such as firewall. Hackers can easily break into wireless routers and other devices, then control them and access other apparatus to steal private information.

6.2.6.2 Emerging Threats to AI Cybersecurity

In 2019, AI applications such as "face change" and "sound altering" apps induced new cybersecurity panic. In early 2019, Hollywood actress Scarlett Johansson was

[5] *5G Network Security Research Report* released by Qihoo 360 Technologies, May 6, 2019.

[6] The Current Security Status and Existing Security Risk Analysis of Internet of Things [OL], April 28, 2019.

subjected to face change on DeepFakes and other software. The "AI face change" incident resulted in panic and question about the safety of artificial intelligence technology in society, and led to other attempts at AI face change. Attackers can use AI to simulate real user's voices, images or behavior patterns, implement more accurate automated spear phishing, create smart botnets with AI malware, and conduct deep penetration and attacks on critical infrastructure, which greatly amplifies the impact of attacks. According to Forbes, FaceApp developed by Wireless Labs, a Russia company, had been downloaded by over 100 million users and recorded more than 150 million faces. Should it be used for political purposes or cybercrimes, the consequences would be unimaginable.

6.2.6.3 Blockchain Business Faces Serious Cybersecurity Problem

The blockchain and distributed ledger technology provide new solution to trust issues in cyberspace, yet the technology itself still faces many challenges in terms of cryptographic algorithm security, protocol security, use security, system security, etc. Its blockchain business security system, in particular, is underdeveloped, making it a favored choice for attacks and accounting for about 80% of losses in total. According to *2018 Blockchain Safety Report* released by BCSEC and PeckShield, in 2018, economic loss caused by security incidents in blockchain applications reached \$2.238 billion, representing a 253% increase from 2017. There were 138 security incidents in 2018 compared with 15 in 2017. The prevalence of security issues affects confidence and experience across the industrial community, and risks related to this field call for attention.

6.3 Countries Actively Building Cybersecurity Protection Capability

In 2019, countries around the world continue to build up their cybersecurity capability. Specifically, they have formulated and improved cybersecurity strategic plans, laws and regulations, stepped up cybersecurity management systems and mechanisms, strengthened security protection of critical information infrastructure, enhanced data protection, developed new technologies in cybersecurity protection, accelerated cybersecurity talent cultivation, and deepened international cybersecurity cooperation, to continuously boost their cybersecurity protection capability.

6.3.1 Optimizing Top-Level Cybersecurity Design

Countries around the world have further strengthened their strategic planning for national cybersecurity and issued strategic reports, policies and laws related to this field. In September 2018, the United Arab Emirates issued a national cybersecurity strategy to ensure information and communication security in five aspects, including establishing a comprehensive response system to deal with underlying cybersecurity risks and training cybersecurity talents. In November 2018, South Africa Parliament formally passed *Cybercrimes and Cybersecurity Bill*, which aimed to align cyber laws of South Africa with those adopted by other countries in response to the growing cybercrimes and other cybersecurity threats. In December 2018, Egypt officially launched *National Cybersecurity Strategy 2017–2021* to deal with cyber threats such as IT infrastructure intrusion and destruction, cyber terrorism and cyber warfare. In response to new cybersecurity environment after Brexit, the United Kingdom began to build its cybersecurity strategy system. In December 2018, it released *National Cybersecurity Capability Primary Strategy: Improving the UK's Cybersecurity Capabilities* to promote and strengthen its cybersecurity competence. In January 2019, Vietnam updated its *Law on Cybersecurity*, stipulating the protection of national critical information infrastructure, emergency response to cybersecurity incidents and other matters. In April 2019, South Korea issued *National Cybersecurity Strategy Guidelines*, in which it proposed six strategic projects including improving countermeasures to cyberattacks, and decided to establish a national information sharing system and build a cybersecurity defense line. In June 2019, EU's *Cybersecurity Act* came into effect, which included stipulations on strengthening the cybersecurity structure, the control of digital technology, and the fulfillment of cybersecurity obligations. It established groundwork for the subsequent *E-Privacy Regulation* and *E-Evidence Regulation.*

6.3.2 Gradually Improving Cybersecurity Systems and Institutions

6.3.2.1 Establishing Cybersecurity-Related Institutions

Many countries have established high-level specialized cybersecurity management institutions to strengthen overall coordination, supervision and management of cybersecurity work. In October 2018, India approved the establishment of Ministry of National Network Defense, Ministry of National Space Defense and Special Operations Command to respond to a trinity of emerging threats from space, cyberspace and special operations. In March 2019, U.S. Department of Defense and Department of Homeland Security announced the establishment of Cyber Protection and Defense Promotion Team, hoping to improve American government's ability to respond to cyber threats with the support of senior management. EU's *Cybersecurity*

Act designated European Network and Information Security Agency (ENISA) to be permanently in charge of EU's cybersecurity to further enhance cybersecurity within EU.

6.3.2.2 Promoting Co-establishment and Sharing of Cybersecurity Monitoring and Early Warning Systems

Countries have stepped up the construction of cybersecurity perception, threat monitoring and early warning systems in support of the detection, protection, response and recovery related to cybersecurity. Australia advocates the sharing of threat information between government and businesses. In August 2018, National Cybersecurity Center was newly established and Cyber.gov.au was updated to help Australian people and businesses to better avoid cyber threats and attacks. In January 2019, the United States issued *2019 National Intelligence Strategy*, which proposed mission objectives, such as cyber threat intelligence and information sharing and maintenance. U.S. Department of Homeland Security also updated its cyber information sharing system accordingly and requested the adoption of automatic threat intelligence sharing technology in over 200 enterprises, organizations and government departments. In early 2019, the Philippines announced the establishment of Cybersecurity Management System Project (CMSP), a national cyber threat intelligence sharing system, to enhance its ability to monitor and respond to cyberattacks.

6.3.3 Strengthening Security Protection of Critical Information Infrastructure

Countries pay great attention to the security of critical information infrastructure and continuously promote the security protection thereof by improving management, establishing relevant certification system, and strengthening attack information sharing. In November 2018, the United States passed *Cybersecurity and Infrastructure Security Agency Act of 2018*, which stipulated the establishment of Cybersecurity and Infrastructure Security Agency (CISA) under U.S. Department of Homeland Security and promoted the management of cybersecurity affairs to federal level. The Act meant that U.S. government made the security of critical infrastructure as a core part of its national security by clarifying the according management system. EU's *Cybersecurity Act* proposed to develop the first EU-wide cybersecurity certification program, and EU member states established informal agreements on the certification of key infrastructure such as energy, water and power, and banking systems. European Commission issued *Recommendation on 5G Cybersecurity*. It required member states to carefully assess 5G network infrastructure and to take into full account of technological risks and risks related to the behaviors of suppliers or operators (including operators from third countries), to ensure the security of

5G network. As the host country of the Tokyo 2020 Olympic Games, Japan took the opportunity to strengthen its collaboration with enterprises and citizens to share information on cyberattacks at home and abroad among all parties, and to improve its response to cyberattacks on important infrastructure such as utility and airports. In addition, South Africa specifically introduced *Critical Infrastructure Protection Act*. The Philippines formulated *National Cybersecurity Plan for 2022*, which focused on the protection of national network and critical infrastructure. Canada also passed *Critical Infrastructure Cybersecurity Guidelines.*

6.3.4 Speeding Up the Protection of Data Security and Personal Information

EU's *General Data Protection Regulation* (GDPR) has set a great example. Countries actively formulate and improve relevant policies and laws, and to strengthen the protection of data security and personal privacy becomes a consensus. In July 2018, India promulgated the first personal data protection law *2018 Personal Data Protection Bill (Draft)*. The Central Bank of India implemented new regulations with great resolution, demanding that its consumers' payment data be stored in India for monitoring, inspection and access. In June 2019, Egypt passed its first data protection regulation. In July 2019, Brazilian Senate passed a proposal to incorporate the protection of personal data on digital platforms into its Constitution as a fundamental right of citizens. EU countries have formulated or revised laws and regulations in alignment with GDPR. In order to meet the requirements of GDPR, Belgium issued *Law on the Protection of Natural Persons regarding the Processing of Personal Data on July 30, 2018*. In December 2019, Spain passed *Organic Law 3/2018, of December the 5th, On the Protection of Personal Data and the Guarantee of Digital Rights*, which largely invoked the relevant provisions in GDPR. In January 2019, Finland's new *Data Protection Law* came into effect, which recognized the legal effect of GDPR in the form of domestic law.

6.3.5 Accelerating Deployment of Emerging Technologies in Cybersecurity

6.3.5.1 Countries Actively Promote the Sound Development of Artificial Intelligence

At present, about 30 member states of the United Nations have formulated national AI development strategies, which emphasize safety as a priority throughout the entire life cycle of artificial intelligence and give full play to AI's strategic value in building up national cyber defense capability. In February 2019, the United States launched

American Artificial Intelligence Initiative in recognition of AI's importance in the field of traditional security. The initiative was designed to maintain American leadership in the field of artificial intelligence. *2018 Summary of Artificial Intelligence Strategy of Department of Defense* that was released later on further advocated the establishment of AI solutions in America's global defense system, an interoperable solution that could be shared among its allies and partners. In April 2019, EU issued *Ethics Guidelines for Trustworthy AI*, which proposed that future AI systems should meet seven requirements including robustness and security, privacy and data management, and transparency. In May 2019, Organization for Economic Cooperation and Development (OECD) passed the first set of intergovernmental policy guidelines on artificial intelligence to ensure the design of AI systems met impartial, safe, fair and trustworthy international standards.

6.3.5.2 Blockchain's Natural Cyber Defense Attribute Contributing to Its Deployment Around the World

Featured by decentralization, anonymity and tamperproof, blockchain can effectively improve the security of protection mechanisms such as encryption and authentication, and enhance cybersecurity protection capability. Governments of various countries have promulgated a series of laws, policies and plans to guide and regulate the development of blockchain industry. In June 2018, a research lab under Russian Ministry of Defense announced plans to develop blockchain ledger system to enhance the cybersecurity certification capability of its military network. In November 2018, U.S. Defense Advanced Research Projects Agency (DARPA) announced Applications and Challenges of Consensus Protocol Program, which was designed to study the safe application of distributed consensus protocol technology in key data storage and computing tasks. In addition, Space & Terrestrial Communications Directorate (S&TCD) under U.S. Army tried to use blockchain to check for discrepancies in its communication data and cybersecurity issues. In May 2019, both Singapore and Canada announced the completion of their cross-border digital currency payment test, which aimed to improve the safety and timeliness of payment transactions through distributed ledger technology.

6.3.5.3 Advantages of Quantum Communication in Cybersecurity Attracting Attention Around the World

Compared with traditional communication methods, quantum communication embodies more safety, transforms traditional communication and calculation methods, and ensures secured exchange of information. In October 2018, EU officially launched the 10-year Quantum Technologies Flagship Program with a total investment of €1 billion, which was expected to deliver a Pan-European Quantum Security Internet around 2035. In December 2018, the United States passed *National Quantum Initiative Act*, which established a 10-year National Quantum

Initiative and three organizations: National Quantum Coordination Office, Subcommittee on Quantum Information Science, and National Quantum Initiative Advisory Committee, and aimed to maintain America's global leadership in the field of quantum communication. The United Kingdom will build a national quantum communication network in the next 10 years, and South Korea plans to build a secured national quantum communication test network in three stages by 2020.

6.3.6 *Speeding Up Cybersecurity Talent Training*

Under the grim outlook of global cybersecurity, there emerges great need for cybersecurity talents, presenting a huge gap between the supply and demand. According to a report released by U.S. Information Security Certification (ISC) in October 2018, there was a 2.93 million gap in global cybersecurity talents, with the Asia–Pacific region taking up most of the gap, in need of 2.14 million talents.[7] Countries all over the world attach great importance to the cultivation of cybersecurity talents, and implement various strategies to strengthen talent training, to improve the competence of cybersecurity practitioners, and to supply talents to maintain cybersecurity.

6.3.6.1 Broadening Cybersecurity Talent Recruitment Channels

In April 2019, eleven federal agencies including U.S. Central Intelligence Agency and Ministry of Defense, launched Cybersecurity Talent Program to provide support to college students who applied for cybersecurity internships. The British government planned to double the number of visas for the first category (excellent talents) in order to attract highly-skilled foreign talents in cybersecurity and other fields. The United States issued *Executive Order on America's Cybersecurity Workforce* in May 2019, which required to take multiple measures to build up its cyber talent team. U.S. Department of Digital Services (DDS) stated that U.S. Department of Defense developed a plan to recruit scientific and technological talents with the help of private human resources companies, and to attract groups that traditional recruitment methods failed to cover.

6.3.6.2 Attaching Importance to Knowledge Training and Skill Education on Cybersecurity

In September 2018, an American congresswoman put forward *Cyber Ready Workforce Act* to the House of Representatives, which would establish a grant program within Department of Labor to support the creation, implementation, and expansion of registered apprenticeship programs in cybersecurity. In November, Cisco

[7]Data Source: *2018 Cybersecurity Talent Report* released by ISC in October 2018.

worked with the British police, helping them master basic cybersecurity knowledge. Singapore established Armed Forces Cyber Defense Academy to provide training to cybersecurity personnel at Ministry of Defense and the Armed Forces.

6.3.6.3 Governments and Universities Teaming Up to Train Cybersecurity Talents

Government agencies are teaming up with universities to cope with the shortage of cybersecurity skills and talents. In March 2019, U.S. Senate introduced *Cyber Security Exchange Act* to promote cybersecurity talent exchange. Federal Bureau of Investigation (FBI) and the University of North Georgia have established a cooperation to better assess cyber threats and prevent potential attacks. The North Atlantic Treaty Organization (NATO) works with Concordia University of Canada, using its expertise to carry out research projects, improve cybersecurity measures, and form an international cybersecurity team. Government of Wales in the United Kingdom invested approximately $13 million in new National Digital Center that is operated by the University of South Wales, to provide cybersecurity technological training.

6.3.7 Expansion of International Cyberspace Cooperation

6.3.7.1 Deepening Cybersecurity Communication and Cooperation Between Countries and Regions

In October 2018, Uruguay and 20 European Commission member states signed *Convention for the Protection of Individuals with regard to Automatic Processing of Personal Data* to strengthen the international protection of personal data. In November 2018, French president Macron initiated *Paris Call for Trust and Security in Cyberspace* and put forward a series of proposals to enhance trust, security and stability in cyberspace. 224 companies and 92 non-profit organizations from 51 countries including all EU member states signed the agreement. In January 2019, EU-Japan Data Exchange Agreement came into effect. The agreement enables free data transmission between EU and Japan. Companies in EU member states are given access to the personal data of 127 million Japanese consumers.

6.3.7.2 Further Expansion of Multilateral Cybersecurity Cooperation Across Different Platforms

The United Nations, the Group of 20 (G20), Shanghai Cooperation Organization (SCO), the BRICS, the Asia–Pacific Economic Cooperation Organization (APEC) and other multilateral frameworks actively carry out cybersecurity cooperation. In September 2018, International Telecommunication Union (ITU), private sector, academia, and non-governmental organizations jointly issued a national cybersecurity strategy guideline, which aimed to help countries formulate and implement national cybersecurity strategies. In November 2018, Global Commission on the Stability of Cyberspace (GCSC) released six global regulations to enhance the security and stability of global cyberspace. In December 2018, the United Nations General Assembly passed two resolutions, *Developments in the Fields of Information and Telecommunications in the Context of International Security* and *Countering the Use of Information and Communications Technologies for Criminal Purposes*, which aimed to protect the rights of all countries in the field of cybersecurity.

6.3.7.3 Further Expansion of Cooperation on Combating Cybercrime

In September 2018, the United Nations Office on Drugs and Crime (UNDOC), the Group of 77 (G77) and Russia jointly organized an informal meeting on the prevention and combat of cybercrime at the United Nations headquarters in Vienna. The meeting focused on topics such as international cooperation on combating cybercrime and government-enterprise collaboration to fight cybercrime. In March 2019, the 5th meeting of the United Nations Group of Governmental Experts (UNGGE) on Cybercrime was held in Vienna. Countries focused the practical discussions on two topics of cybercrime, "law enforcement and investigation" and "electronic evidence and criminal justice", and put forward dozens of guidelines on specific rules. In May 2019, the United States and European police announced in Hague, the Netherlands, that the law enforcement agencies of the United States, Georgia, Ukraine, Germany, Bulgaria, and Moldova worked together and took down a complex cyberfraud ring based in Eastern Europe, which targeted small companies and charitable organizations with phishing emails to obtain information on the victims' online bank accounts and stole approximately $100 million in deposit. In July 2019, Interpol and cybersecurity company Kaspersky signed a new five-year agreement to strengthen cooperation on combating global cybercrime.

6.4 Continuous Development of Global Cybersecurity Industry

Countries around the world have optimized policies and measures and increased investment to accelerate the development of cybersecurity industry and promote the continuous growth of cybersecurity companies.

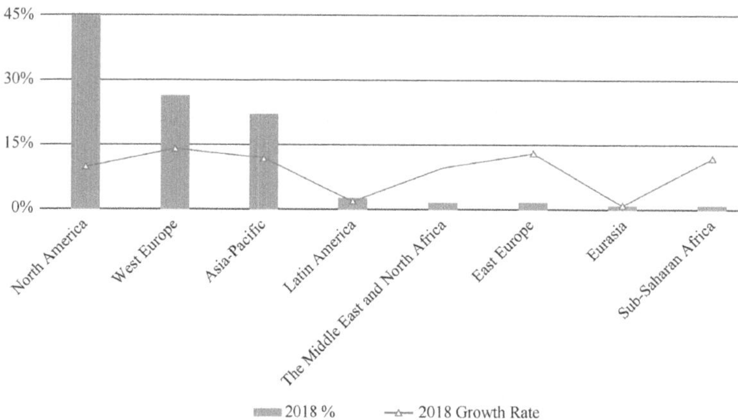

Fig. 6.1 Regional distribution and growth of global cybersecurity industry in 2018

6.4.1 Growing Cybersecurity Industry

In 2018, global cybersecurity industry reached $112 billion, representing a 11.3% increase year on year and a record high since 2016, and was expected to reach $121.7 billion in 2019.[8] In terms of regional distribution, North America still claimed the largest share of global cybersecurity market, followed by Western Europe and the Asia–Pacific region. Specifically, cybersecurity industry in North America, mainly in the United States and Canada, was $50 billion in 2018, representing a 10% increase over 2017 and accounting for 45% of global cybersecurity market. Cybersecurity industry in Western Europe, headed by the United Kingdom and Germany, reached $29.4 billion, representing a 14% increase over 2017, with the fastest growth around the world. Cybersecurity industry in the Asia–Pacific region, led by China, Japan and Australia, reached $24.6 billion in 2018, representing a 12% increase over 2017. Cybersecurity industry in the Middle East, Eastern Europe, Latin America and other regions totaled $8 billion, accounting for only 7.18% of global market share. See Fig. 6.1 for regional distribution and growth of global cybersecurity industry in 2018.

6.4.2 New Technologies and Applications Empowering Industrial Development

As the integration of emerging technologies such as AI and big data continues to accelerate, new AI applications spring up. New business form of AI-aided network offense and defense has initiated substantive deployment, which promotes the rapid development of global cybersecurity industry, one of the most active fields of AI

[8]Gartner, Information Security and Risk Management, Worldwide, 2017–2023, 2019 Q 1 Update.

application. According to the statistics of CB Insights, there are more than 80 AI-aided security companies in the world. Among them, Tanium, an automated terminal protection provider, and Cylance, an intelligent predictive analysis provider, both have a market value of over $1 billion. Many innovative tech companies such as Skycure, Darktrace, Authbase and CyberFog actively promote the application of artificial intelligence to the fields of identity management, cyber fraud protection, abnormal behavior analysis, mobile security, and IoT security.

6.4.3 Steady Development of Listed Cybersecurity Companies

As innovation on network information technology speeds up, traditional cybersecurity protection systems are no longer applicable and large cybersecurity firms find it particularly hard to make the transition from existing models. On the other hand, emerging cybersecurity technologies and products constitute a stronger drive for the market and cybersecurity start-ups gain momentum based on their technological advantages. In terms of R&D investment, there has been rapid growth among listed cybersecurity companies. In 2018, the average R&D investment of major listed cybersecurity companies was $316 million, an 11% increase from $285 million in 2017. R&D investment grew by 21% on average, maintaining at a high level. In terms of operating income, global listed cybersecurity companies maintained rapid growth in 2018. According to financial reports of listed companies, ten typical cybersecurity companies, including CheckPoint, Symantec, Palo Alto Networks, and Trend Micro averaged $1.652 billion in operating income, registering a 13% increase from 2017.

6.5 Growing Militarization of Global Cyberspace

At present, international landscape undergoes profound changes, and the struggle between major powers has complicated geopolitical factors even more. International competition in cyberspace is intensifying, and global struggle for network control is increasingly fierce. Countries around the world take cyberspace as a national strategic priority and a high ground worth competing for. They step up strategic deployment of cybersecurity and build up cyber defense capability. Certain countries strengthen their cyber deterrence strategies and continue to step up cyber warfare preparations and cyber force building, in an attempt to ensure absolute security against any risks. These efforts have fueled the cyberspace arms race, intensified global cyberspace militarization and posed new challenges to world peace.

6.5.1 A Surge of Cyberspace Strategies Changing Traditional Warfare Rules

In order to consolidate its advantage in cyberspace, the United States has issued strategic reports and policy bills such as *National Cyber Strategy* and *Cybersecurity Strategy for Department of Defense*, taking cyberspace into the main battlefield of international competition. It also proposes to apply traditional laws of war such as *Law of Armed Conflict* to cyberspace, and continues to expand the applicable scope of regulations to include digital weapons for national defense. *National Cyber Strategy 2018* puts forward the concept "preemptive defense" as a policy support for the United States to implement preemptive cyber strikes. The United States mentions in the latest *Nuclear Posture Review* that it could use nuclear weapons to respond to non-nuclear strikes such as cyberattacks. In addition, reports state that U.S. government has further reduced restrictions on cyberattacks, allowing its cyber forces to attack opponents more frequently and easily. In December 2018, Japanese government updated *National Defense Program Outline*, a framework document for developing national defense in the next 10 years, and for the first time proposed the concept of "cross-domain defense". The proposal included building joint defense force involving space, network and other fields, and put its focus on strengthening cyber warfare rapid response capability and counterattack capability. In January 2019, French Defense Ministry issued an offensive cyber warfare doctrine that combined traditional military operations with cyber warfare, and provided military operations with cyber advantages to achieve mission objectives. In May 2019, Russian president signed *Sovereign Internet Law*, which devised the development of an independent network in Russia that could operate stably and ensure network connection within the country should the outer world suffer Internet crash or other attacks.

6.5.2 Large-Scale Cyber Force Intensifying Risk of War

Globally, more than 100 countries have set up cyber warfare units. The risks of cyber warfare between countries rise and pose a greater threat to the strategic stability of cyberspace. The United States upgrades its Cyber Command to a top-level command, and its 133 cyber task forces already have combat capability. In addition, its navy, army and air force are strengthened in cyber defense capability. In April 2019, U.S. Army established a new multi-domain combat force that combined cyber warfare, intelligence and strike. In January 2019, NATO established a new cyber command that would go fully operational in 2023 in order to monitor cyberspace in a timely manner and effectively counter various cyber threats. Japan proposed in its new *National Defense Program Outline* to expand its cyber defense force and establish a new unit under Ground Central Command of Japan Ground Self-Defense Force (JGSDF). In September, French President signed *French Military Planning Act of 2019–2025*, which stipulated that French government would add thousands of operators to its cyber troops by 2025 to improve its cyber warfare capability. *Daily Star* stated that British Special Air Service Regiment established a cyber warfare unit to counter foreign cyberattacks and assist battlefield operations.

6.5.3 Cyber Warfare as an Important Strategic Deterrence Between Countries

Since 2019, the United States and many European nations have held exercises frequently, including "Saber Guardian 2019", "Cyber Flag 19.1", "Cyber Lightning 2019", "2019 Network X-Game", "Blue OLEx 2019", "Exercise Mercury", and "Locked Shield 2019", with growing participation, expanding fields and more complicated simulation environment. The United States has established a network information operations development center and a network research and analysis lab to keep up with changes in the battlefield and provide rapid support to cyber operators on mission. In July 2018, U.S. Department of Defense sought to develop an advanced cyber weapon system, "United Platform", which could assist cyber forces in acquiring defensive and offensive tools. In August 2018, Defense Advanced Research Projects Agency (DARPA) commissioned Packet Forensics, a cybersecurity company, to develop a botnet identification system that could automatically identify and locate hidden botnets. Internet warfare is coming. Several U.S. media reported that U.S. government had U.S. Cyber Command carry out a retaliatory cyberattack on Iran in June 2019, targeting Iran's computer systems that controlled rocket and missile launches. According to the report, the U.S.-led cyberattack "paralyzed" Iran's weapon system and destroyed a database that was very important to Tehran.

Nowadays, security and stability of cyberspace receives global concern and relates to the sovereignty, security, development and welfare of all countries. In cybersecurity, old problems remain unsolved and new ones emerge. Cyberspace security faces rising uncertainty and complexity. The growing trend of militarization of cyberspace, in particular, poses major challenges to world peace and security. We must understand that cyberspace is interconnected, deeply intertwined with the welfare of all countries. A safe, stable and prosperous cyberspace is of great significance to all countries and the world. Cyberspace should not be a battlefield of competition for countries around the world or a hotbed for crimes. Faced with increasingly complex cybersecurity environment, all countries should strengthen strategic mutual trust and step up communication and cooperation. Based on full respect for other nations' security, countries should work together to combat illegal and criminal activities, enhance global cybersecurity protection capability, reduce cyberspace confrontation, and establish international rules and mechanisms to contain cyber warfare, resist cyber arms race, and build a more peaceful, secure and stable cyberspace.

Chapter 7
International Cyberspace Governance

7.1 Outline

The rapid development of world Internet has given rise to new problems and challenges to traditional political, economic and governance structure. International rules on cyberspace have yet to form. Cyber surveillance, cyberattacks, cyber terrorism, violation of privacy, infringement on intellectual property rights and other illegal and criminal acts afflict the world, calling for immediate cooperation within international community to improve international cyberspace governance and maintain peace and security in cyberspace.

International cyberspace governance stands at a stage of multilateral and overlapping management. Countries around the world pay close attention to international cyberspace governance and strive to promote the formation of universally accepted international rules on cyberspace. Now, driving the reform of global cyberspace governance system has become a growing consensus among international community. State actor serves as an important force in improving cyberspace governance, and relations between major powers become the key to international cyberspace governance. Although affected by changes in the international landscape and the world order, international cyberspace governance has made progress to some extent. In terms of state actors, countries around the world have further devised and improved cyberspace governance propositions, issued laws to regulate cyberspace development, stimulated the development of cyberspace, and enhanced cybersecurity protection capability. The United States and Russia urge the United Nations to keep formulating codes of conduct for cyberspace that are generally recognized by all countries. France initiates *Paris Call for Trust and Security in Cyberspace*, which proposes a new path for cyberspace governance. Japan launches a new framework for cross-border data flow. In terms of non-state actors, international organizations, represented by the United Nations, have vigorously promoted international cyberspace governance and paid more attention to the role of cyberspace governance forums and platforms. International organizations have also reconvened the UN Group of Governmental Experts on Information Security,

Chinese Academy of Cyberspace Studies, *World Internet Development Report 2019*, https://doi.org/10.1007/978-981-33-6938-2_7

created a new UN open-ended working group for information security, and focused on making international cyberspace rules. International cyberspace governance platforms such as International Telecommunication Union (ITU) and Internet Corporation for Assigned Names and Numbers (ICANN) push for continuous progress in formulating technological specifications, bridging the digital divide, and studying international rules based on their own advantages. Traditional international organizations such as the Group of 20 (G20), Shanghai Cooperation Organization, the BRICS, Asia Pacific Economic Cooperation (APEC), and Organization for Economic Cooperation and Development (OECD) set their eyes on cyber issues. For example, member states of Shanghai Cooperation Organization deepen their cooperation in combating cyber terrorism. At the G20 summit, many countries agreed upon cross-border data flow, principles for artificial intelligence and other topics.

Chinese president Xi Jinping attaches great importance to international cyberspace governance and expounds the concept of international governance on many important international occasions, including the "Four Principles", "Five Proposals" and "Four Commons" of global Internet development and governance in particular. The proposals have received broad recognition among many developing countries, and important concepts such as cyber sovereignty and a community with a shared future in cyberspace have gained the public's support. China actively promotes international governance, exchange and cooperation in cyberspace and clarifies its propositions on digital economy and cybersecurity. It also proposes that countries should deepen their pragmatic cooperation based on common progress and win-win outcome, build a bridge of mutual trust and common governance, create and invigorate a community with a shared future in cyberspace, and push global Internet governance system onto a fairer and more reasonable path.

7.2 International Cyberspace Governance Faces an Important Turn

At present, there presents growing fragility and uncertainty with international rules on cyberspace, and various actors actively push governance forward and make certain progress. Meanwhile, the construction of international order in cyberspace faces a series of challenges. The competition among major powers in cyberspace intensifies, calling for growing consensus on cyberspace governance.

7.2.1 Emerging Fragility and Uncertainty of International Cyberspace Rules

So far, governments, international organizations, technological communities and enterprises have been the main driving force of international cyberspace governance.

As the international landscape enters a stage of restructuring and transformation, the struggle between major powers in the physical world is directly projected into cyberspace, and some countries even use Internet as an important means to undermine and contain other countries. As a long-standing power in international governance, the United States starts to adjust its global strategy on a large scale and reduces its investment in international cyberspace governance. New technologies and applications add new elements to the game in cyberspace. The governance models of multilateralism and multi-stakeholder are merging and have yet to form synergy in international cyberspace governance. With the restructuring of international landscape and struggle between major powers, rules on international cyberspace governance weaken or even sop to work. International cyberspace governance faces great uncertainty.

7.2.2 Various Participants Urgently Needed in International Cyberspace Governance

There is a growing need for state actors and non-state actors to participate in e international cyberspace governance. In the context of national strategy, countries promote the development of new technologies and applications such as artificial intelligence, 5G and big data, and pay attention to cross-border data flow and data security issues, hoping to establish international rules for new technologies and applications in line with their own interests and development. The struggle between major powers in cyberspace has intensified, which directly affects corporate activities and other non-state actors in cyberspace. Companies may face risks caused by the adjustments in national policy such as supply chain security, data protection, intellectual property rights, and market access. More and more companies take part in the formulation of international rules on cyberspace. An example is that Microsoft proposes *Digital Geneva Convention*. At the end of 2018, the United Nations reconvened its Group of Governmental Experts on Information Security and established a new open-ended working group, making it easier for member states to participate more widely in the formulation of international cyberspace rules and strengthening the United Nations' authority in international cyberspace governance. In addition, traditional international organizations such as G20, Shanghai Cooperation Organization, the BRICS, and the Asia-Pacific Economic Cooperation (APEC) also actively advocate and advise on digital economy, combating cyber crime and cyber terrorism and other fields.

7.2.3 *Slow Progress on International Cyberspace Rules*

The rapid development of new technologies and applications has led to the rise of new governance issues and institutional needs. International community is exploring governance rules in related fields, but the overall progress is relatively slow. Presently, cyberspace security faces grim outlook. Countries are more concerned about their own security than contributing to the formulation of international rules. The struggle among major powers has intensified, casting a shadow over the formulation of relevant rules. Besides, forces with vested interests in the current cyberspace governance are protective of their own interests and reluctant to make adjustments and changes, while other willing forces lack the ability to set the agenda and lead the change. As a result, the promotion of international rules on cyberspace is generally in a stalemate.

7.2.4 *Major Challenges in the Construction of International Cyberspace Governance Order*

At present, countries shift their focus on cybersecurity from international security to their own security. The United States, in particular, has adjusted its global strategy and prioritized its own needs, unwilling to assume more responsibility in providing public products. The shift has made it difficult for international coordination mechanisms to function and hindered the progress of international collaboration. In recent years, there have been frequent interactions between the United States, Europe and Japan. They have strengthened cooperation in the fields of cybersecurity and digital economy and promoted the establishment of related regulatory systems through aligned regulatory stance, economic partnerships, data agreements and other measures. China and Russia exert their influence through UN framework, Shanghai Cooperation Organization and other mechanisms. These regional or bilateral mechanisms are important channels and useful supplements for advancing the overall progress of international cyberspace governance. However, it should be noted that there are certain disagreement and even conflicts among these governance mechanisms and rule systems dominated by different countries. There is obvious fragmentation with cyberspace governance platforms, governance plans and governance practices, which has caused widespread concern in international community.

7.3 Continuous Progress of International Cyberspace Governance Platform

International governance platforms exemplified by the United Nations continue to promote international cyberspace governance, and achieve positive progress in the development of cyberspace rules, digital economy and cybersecurity protection.

7.3.1 The United Nations Internet Governance Forum (IGF)

The United Nations increasingly values Internet Governance Forum (IGF) and pushes it to assume more responsibility in the field of international cyberspace governance, trying to build it into a platform for stakeholders to exchange information and share best practices on Internet-and-tech-related policies. Countries also pay more attention to IGF and view it as the main platform for promoting the concepts of international cyberspace governance. In November 2018, the 13th Annual Session of UN Internet Governance Forum was held in Paris, France. UN Secretary-General Antonio Guterres, French president Emmanuel Macron and other leaders attended the venue and delivered speeches. In his speech, Guterres called on all parties to convert "digital risks" into "digital opportunities", attach greater importance to network information security, strengthen cooperation in the digital field, narrow the digital gap between developing and developed countries, and increase innovative trusted network solutions. He also emphasized that discussion was not enough for Internet governance, and that we should formulate policies to regulate management rather than leave it to the invisible hand of market forces. At the opening ceremony, French president Macron announced *Paris Call for Trust and Security in Cyberspace*, putting forward a series of propositions on enhancing trust, security and stability in cyberspace, and hoping to create a "third way" for Internet governance. The proposal generated positive influence in international community.

7.3.2 The United Nations High-Level Panel on Digital Cooperation

In 2018, the United Nations established High-Level Panel on Digital Cooperation and achieved positive results in promoting digital cooperation. In June 2019, the panel submitted *The Age of Digital Interdependence* to Secretary-General of the United Nations, which mainly included three parts: "Leaving No One Behind", "Individuals, Societies and Digital Technologies" and "Mechanisms for Global Digital Cooperation". The report calls for building an inclusive digital economy and society where every adult has access to affordable digital networks and digital financial and medical services by 2030. It aims to establish a wide-range multi-stakeholder alliance, achieve the sustainable development, and share "public digital products" and data. It also clearly proposes to forge new global digital cooperation mechanism, and is regarded as a programmatic guide leading the future development of global digital economy.

7.3.3 The United Nations Group of Governmental Experts (UNGGE) and Open-Ended Working Group (OEWG) on Information Security

In 2004, the United Nations General Assembly established the first UN Group of Governmental Experts (UNGGE) on Information Security to discuss international rules for cyberspace. In 2017, the fifth UNGGE, involving 25 countries including the United States, Russia and China, failed to reach a consensus on issues related to the application of international laws to cyberspace and the specific use of codes of conduct for responsible state action, or produce any agreed document, which led to doubts about the necessity of the next GGE among international community. In December 2018, supported by the United States and Russia, the United Nations General Assembly set up the 6th GGE on Information Security and announced the establishment of new Open-End Working Group (OEWG) for in-depth discussions on responsible state action in cyberspace. The open-ended working group was formed in June 2019, which welcomed the participation of all UN member states interested in the matter. Its main task is to further discuss the standards and norms for cyberspace as well as the principles and implementation methods of relevant responsible states, and to study existing and potential information security threats and to come up with measures to build trust and capability. OEWG will also organize intersessional consultative meetings with companies, NGOs and academia to discuss issues like cyberspace regulations. The 6th GGE on Information Security will operate for three years. It will carry on the work started by former GGEs and study feasible measures to deal with threats in the international information security field.

7.3.4 World Summit on the Information Society (WSIS)

World Summit on the Information Society (WSIS) focuses on supporting sustainable development. The summit is an important communication platform under the UN framework for global ICT stakeholders to exchange information, promote knowledge innovation, share practical experience, grasp industry trends and develop partnerships. WSIS was held in Geneva, Switzerland, in April 2019. Under the theme of achieving sustainable development with ICT, the summit aimed to promote and implement ICT technological solutions and action plans in support of the United Nations 2030 Agenda for Sustainable Development, and attracted more than 3,000 ICT experts and action participants. During the summit, the Ministerial Round Table emphasized the importance of WSIS actions, making WSIS technological action items the key framework for realizing the goals of the United Nations Sustainable Development. In addition, WSIS calls on all parties to share scarce resources, strengthen practical cooperation, and establish digital skill and ICT incubation programs to combat cyberattacks and guarantee the safe and secured use of digital technology for the benefit of society.

7.3.5 International Telecommunication Union (ITU)

International Telecommunication Union (ITU) continues to promote the formulation of technological standards, bridge the digital divide, closely follow the development of 5G and artificial intelligence, and study and formulate relevant standards. In July 2019, the Working Party 5D (WP 5D) under ITU's Radiocommunication Bureau held a meeting in Brazil. Over 180 representatives from government authorities, telecommunications manufacturers and operators, and research institutions around the world attended the meeting and discussed 5G technological solutions to prepare for the announcement of 5G technological plan in 2020. In August 2019, ITU officially released an international standard on 5G and artificial intelligence, *Architecture and Framework for the Use of Machine Learning in Future Networks Including 5G.* The standard lays the foundation for "integrating machine learning into 5G systems and future networks in a low-cost but highly effective way". It proposes a set of architectural requirements and specific architectural components that meet the needs of operators, as well as guidelines on how to integrate these components into 5G and future networks and various technology-specific underlying networks. In 2018, at the Plenipotentiary Conference in Dubai, ITU member states approved *Four-Year Strategic and Financial Plan*, which included the development of information and communications technological infrastructure and relevant skills needed to vigorously promote inclusive economic growth, boost innovation and bridge the digital divide.

7.3.6 Internet Corporation for Assigned Names and Numbers (ICANN)

In 2019, Internet Corporation for Assigned Names and Numbers (ICANN) formulated a new five-year strategic plan, to keep pushing the reform, promote important domain name policies and technological specifications and actively respond to the impact of data protection. In May, ICANN Board of Directors approved *Five-Year Strategic Plan for Fiscal Years 2021–2025 (Draft).* The Plan states that ICANN's core mission is to ensure the stable and safe operation of Internet's unique identifier system. Its vision is to fulfill its responsibility as an independent, reliable and multi-stakeholder manager of Internet's unique identifier system and provide an open and collaborative environment. ICANN's five major strategic goals are as follows: to strengthen the security of the Domain Name System and the DNS Root Server System, to improve the effectiveness of ICANN's multi-stakeholder model of governance, to evolve the unique identifier systems, to address geopolitical issues impacting ICANN's mission, and to ensure ICANN's long-term financial sustainability. ICANN has also actively promoted the formulation of policies for new-generation domain name lookup system (WHOIS) under the framework of the EU's *General Data Protection Regulation*, and developed a new WHOIS service that meets data protection regulations. ICANN and the root server operating organization community have made preliminary progress on improving the governance structure and mechanism of the root server system, and proposed a concept document, which aimed to establish a more transparent, stable and sustainable root server system governance model. There has been somewhat limited progress with strengthening ICANN's accountability and pushing its transparency reform, but motivation is lacking for further development. It is necessary to step up negotiation and find common ground for all parties around the world.

7.3.7 Internet Society (ISOC)

Internet Society (ISOC) actively advocates the promotion and development of Internet and supports the promotion of routing security, IoT security, privacy protection and other related processes. In 2019, under the theme of "Connecting the world, Improving technological security, Building trust, and Shaping the future of Internet", ISOC developed an interconnection-centered roadmap to improve the security foundation of Internet technology and its service to mankind. ISOC released *2019 Internet Report, Integrating Internet Economy: How Integration Will Affect the Technological Innovation and the Use of Internet*, which further explored the development of Internet economy and examined the growing role of Internet platforms in Internet economy and its possible impact on society, innovation, competition, economy and the broader architecture of Internet.

7.3.8 Global Commission on the Stability of Cyberspace (GCSC)

Global Commission on the Stability of Cyberspace (GCSC) has been committed to studying cyberspace governance norms since its establishment. Cyberspace security rules are the focus of GCSC's discussion. In November 2018, GCSC released six new global regulations that aimed to ensure the rational use of cyberspace and improve the security and stability of cyberspace, stated as below:

(1) Norm to avoid tampering;
(2) Norm against commandeering of ICT devices into botnets;
(3) Form for states to create a vulnerabilities equities process;
(4) Norm to reduce and mitigate significant vulnerabilities;
(5) Norm on basic cyber hygiene as basic foundational defense;
(6) Norm against offensive cyber operations by non-state actors.

GCSC promotes the proposed norms to all parties in various forms and receives the attention of international organizations such as the United Nations and European Union. In January 2019, GCSC held a public seminar that focused on how international law, human rights, Internet governance, development and sustainable development affected cyberspace peace and security, along with recommendations on cyberspace framework for international peace and security in the future.

7.3.9 World Internet Conference

World Internet Conference (Wuzhen Summit), proposed and organized by China, is an annual event in global Internet community. Since the first convention in 2014, it has been held in Wuzhen, Zhejiang, for six consecutive years, and recognized as the most representative international Internet governance platform in China. Over the past six years, World Internet Conference has given full play to its positive role in the interconnection, sharing and co-governance of cyberspace by building first-class platform, deepening exchange and cooperation, and jointly fostering a mutually beneficial and win-win online market. As countries around the world are more closely connected in cyberspace with more frequent exchange and deepening cooperation, cyberspace governance moves onto a fair and reasonable path at a faster pace. In November 2018, under the theme of "Creating a Digital World for Mutual Trust and Collective Governance: Towards a Community with a Shared Future in Cyberspace", the 5th World Internet Conference attracted representatives from governments, international organizations, tech communities, enterprises, and civil organizations to expound on topics such as international cyberspace governance, digital economy, cybersecurity, and cyber technology innovation, in the hope to contribute new wisdom and drive to international cyberspace governance.

7.4 Traditional International Organizations Accelerating Their Participation in International Cyberspace Governance

In 2019, traditional international organizations actively responded to the new trend in international cyberspace governance, urged countries to strengthen exchange and cooperation, and made positive progress in the formulation of international cyberspace rules.

7.4.1 G20

Digital economy has become a focus of the Group of 20 (G20) summit. In June 2019, the G20 summit was held in Japan, focusing on topics such as data governance and artificial intelligence. During the summit, leaders of 24 countries and regions, including China, the United States, Japan, and Germany, jointly signed *Osaka Declaration on Digital Economy* (hereinafter referred to as *The Declaration*). *The Declaration* stated that as digitization changed all fields of economy and society, and data became an increasingly important resource for economic growth, the effective use of data benefited and contributed to the societies of all countries. National and international policy discussions were essential to fully realize the potential of data and digital economy. And countries emphasized that they would actively participate in international policy discussions to fully release the potential of data and digital economy. In addition, *The Declaration* also announced the official launch of the "Osaka Track" for countries to discuss data circulation rules. *G20 Leaders' Osaka Declaration* pointed out that it was necessary to further promote the free flow of data and enhance the trust between consumers and enterprises under the framework of domestic and international laws, and encourage interoperability between different frameworks. *The Declaration* also proposed the establishment of "human-oriented AI" for the first time, along with inclusive growth, sustainable development and well-being, human-oriented values and fairness, transparency and interpretability, robustness, safety and security, accountability and other principles. In addition, the summit also issued *G20 Osaka Leaders' Statement on Preventing Exploitation of the Internet for Terrorism and Violent Extremism Conducive to Terrorism*, emphasizing that protecting citizens' safety and combating terrorist forces were important responsibilities of governments, and all countries were obligated to strengthen cooperation between cyber platforms, review cyber content, and prevent terrorists from abusing cyberspace for terrorist activities.

7.4.2 The BRICS

The BRICS, an important multilateral cooperation mechanism led by emerging economies and developing countries, pays close attention to the topic of digital transformation, carries out the fight against cyber terrorism, and makes meaningful progress in the field of international cyberspace governance. In June 2019, leaders of the BRICS held an informal meeting in Osaka, Japan, and issued *Press Communique of the BRICS Leaders' Informal Meeting*, reiterating the responsibility of all countries to prevent financing terrorist networks and terrorist acts on their own territory, stating again their commitment to combating cyber terrorism, and calling on tech companies to cooperate with the government under legal framework to inhibit terrorists' ability to instigate, recruit, promote or carry out terrorist acts on digital platforms. In August 2019, the 5th BRICS Communications Ministers' Meeting was held in Brazil with in-depth discussions on issues such as the focus of ICT policies, government-enterprise collaboration, strengthening BRICS cooperation under multilateral framework, and promoting digital transformation. The meeting encouraged cooperation among various parties, deepened practical collaboration in multiple fields such as infrastructure interconnection and integration, digital technological innovation, digital transformation and governance, and agreed on building a digital BRICS task force.

7.4.3 Asia-Pacific Economic Cooperation (APEC)

Asia-Pacific Economic Cooperation (APEC) strives to keep up with the trend of the times, and explores new areas of digital economy in search of new drives for growth. In November 2018, the 26th APEC Leaders' Informal Meeting was held in Papua New Guinea. Taking "To Harness Inclusive Opportunities and Embrace a Digital Future" as the theme, the meeting firstly centered on and held discussion of digital economy. Participating leaders echoed positively in the discussion, exchanged their insights on digital economy and digital inclusion, reviewed the course of cooperation, and discussed their vision for the Asia-Pacific region. Among them, president Xi Jinping explained his vision of digital economy and confirmed that digital economy was the future path for the Asia-Pacific region and even the world. We should firmly seize the trend of innovation and development, implement *Internet and Digital Economy Roadmap* in a comprehensive and balanced manner, and fully tap into the growth potential of digital economy. In addition, we should strengthen digital infrastructure and capacity building, boost accessibility of digital economy, and narrow the digital divide, so that members at different development stages can share the fruits of digital economy. This takes all areas of the Asia-Pacific region onto the fast track of economic development.

7.4.4 Shanghai Cooperation Organization (SCO)

Shanghai Cooperation Organization (SCO) encourages member states to deepen cooperation, jointly promote Internet governance according to the case of each state, formulate widely accepted rules, principles and norms for responsible national action in cyberspace, and actively engage in collaborations to ensure regional information security. In May 2019, at a regular meeting of SCO Council of Foreign Minister, foreign ministers of member states called on all UN member states to further promote the development of codes of conduct for responsible state action in cyberspace, and formulate international laws to combat the use of ICT for criminal purposes under the UN's leadership. In June 2019, the Council of Heads of State of SCO held a meeting and signed *Bishkek Declaration of the Council of Heads of State of Shanghai Cooperation Organization*, stating that member states would crack down on the use of ICT to undermine political, economic and social security of SCO member states, combat the spread of terrorism, separatism and extremism through Internet, and oppose any discriminatory practices against the development of digital economy and ICT technology under all circumstances.

7.4.5 *Organization for Economic Cooperation and Development (OECD)*

Organization for Economic Cooperation and Development (OECD) vigorously promotes the formulation of digital economy rules, and advises its member states on the development of digital economy, AI governance, digital taxation and other issues. The recommendations are accepted as guidelines for policy and practice by member states and are increasingly recognized among other international organizations. In May 2019, OECD formally adopted the first intergovernmental policy guidelines on artificial intelligence to ensure that the design of AI systems met impartial, safe, fair and trustworthy international standards. 36 OECD member states along with Argentina, Brazil, Colombia, Costa Rica, Peru, and Romania jointly signed *OECD Principles on Artificial Intelligence* at the Council of Ministers meeting, which were absorbed by the G20 Osaka Summit as the main content of *G20 AI Principles*. OECD actively promotes digital taxation. In March 2019, it held a public consultative meeting in Paris for the public's opinion on the design of taxation rules and technological challenges of digital economy. It strives to form a unified digital economy taxation by 2020, which may change international taxation rules enormously.

7.5 Cyberspace Governance in Some Typical Countries and Regions

In the past year, countries and regions, based on their own conditions and needs, have taken a variety of methods in their exploration of international cyberspace governance and accumulated new experience from their practices.

7.5.1 *The United States*

Since Donald Trump took office, he has paid special attention to maintaining cybersecurity and kept increasing investment in 5G, artificial intelligence, Internet of Things and data governance to ensure America's advantage in cyberspace. In 2019, the United States set its goal of cyberspace governance on suppressing opponents in an all-round way, strengthening cybersecurity and maintaining leadership in cutting-edge technology. In May 2019, Trump signed *Executive Order on Securing the Information and Communications Technology and Services Supply Chain,* forbidding the dealings and uses of foreign information technology and services that would pose a special threat to its national security, foreign policy and national economy in America's view, and publicly picking on Chinese companies. He also lobbied EU, Japan, Australia and other allies into co-signing *Prague 5G Proposal* with the United States, and intended to exclude China from making rules for 5G technology. The United States issued *5G FAST (Facilitate America's Superiority in 5G Technology) Plan* and *National Artificial Intelligence Research and Development Strategic Plan* to ensure America's global leadership in 5G, artificial intelligence and other fields. In addition, the United States continues to strengthen its cybersecurity capacity and increase its budget for cybersecurity, and successively introduces *Securing Energy Infrastructure Act*, *DOD Cloud Strategy*, *National Emergency Communication Plan*, *Corporate Mobile Security Guidelines*, and *Internet of Things Cybersecurity Improvement Act*, to boost America's ability to respond to security risks in all areas of cyberspace.

7.5.2 China

China actively builds Internet governance platform and conducts multi-level, cross-field international cooperation in cyberspace. In November 2018, the 5th World Internet Conference was held in Wuzhen, Zhejiang. The summit achieved fruitful results in terms of exchange of ideas, theoretical innovation, technological display, economic and trade cooperation, and consensus building. In 2019, China further deepened cooperation with EU, Italy, France, the United Kingdom, Germany, Russia, India and other countries and regions in the Internet field. They carried out dialogue and exchanges on topics such as cybersecurity, digital economy, cyber crime, and the development of cutting-edge technologies in cyberspace, and reached broad consensus. China actively promotes the inclusive development of the Asia-Pacific digital economy, implements the "Belt and Road" Initiative, promotes high-end dialogue on the construction of China-ASEAN Information Harbor, Digital Silk Road and Southern Transport Corridor, encourages non-governmental cooperation such as with think tanks and enterprises, and facilitates exchange on cyberspace issues.

China is deeply involved in international cyberspace governance led by important platforms such as the United Nations, G20, SCO, and ICANN. China promotes its concepts on world Internet development and governance, and advocates its ideas on global digital economy and Internet development. At the G20 Osaka Summit, president Xi Jinping profoundly explained China's propositions on the development of digital economy and data governance, and proposed to create a fair, just, and non-discriminatory market environment. He said we could not engage in development behind closed doors or arbitrarily interfere with the market. President Xi Jinping pointed out that we needed to jointly improve data governance rules to ensure the safe and orderly use of data, that we needed to promote the integrated development of digital economy and real economy, strengthen the construction of digital infrastructure, and promote interconnection, and that we needed to boost the inclusiveness of digital economy and bridge the digital divide. President Xi Jinping also emphasized that as a major power in digital economy, China was willing to actively participate in international cooperation and keep an open market to achieve mutual benefit and win-win outcome.

7.5.3 Japan

Japan has further improved its domestic data legislation and proposed Data Free Flow with Trust in the hope to build an international data flow framework. Japan has started amending its *Personal Information Protection Act* to find balance between improving personal information protection and promoting data use. In June 2019, at the G20 Osaka summit, Japan set data governance as a priority and vigorously promoted the establishment of a new international data flow system. As cybersecurity gets increasingly complicated, Japan and the United States have strengthened cooperation in this field. In January 2019, Japan listed six major areas, including cyberspace, as the main content of Joint Operations Planning and Execution System (JOPES) between Japan and the United States, and enhanced strategic coordination and tactical synchronicity between the Self-Defense Force and U.S. military in Japan. In April 2019, the United States and Japan confirmed that for the first time, cybersecurity would be included in the scope of Article 5 of *U.S.–Japan Security Treaty*, and the United States would provide Japan with cybersecurity protection. In addition, Japanese government modeled on the United States and introduced new regulations to limit foreign shareholding in ICT companies on the grounds of maintaining cyberspace security.

7.5.4 European Union

EU continues to promote data security, constantly updates relevant policies and laws, and makes positive progress in data copyrights governance, cybersecurity, and combating misinformation. In May 2019, EU formally implemented *Regulation on the Free Flow of Non-personal Data*, protecting data storage and processing from unfair restrictions across EU by setting up an EU-wide framework. *Regulation on the Free Flow of Non-personal Data* and *General Data Protection Regulation* (GDPR) will ensure the free flow of personal data and non-personal data, promote the construction of a single digital market in the EU, and create a competitive digital economy. In March 2019, EU passed *Directive on Copyrights in the Digital Singles Market* to address the challenges posed by the development of digital technology, strengthen online copyrights supervision, and promote the creation, dissemination and utilization of works in the digital environment. In June 2019, EU formally implemented *EU Cybersecurity Act*, designating European Network and Information Security Agency (ENISA) as a permanent EU institution responsible for better supporting member states in response to cybersecurity threats and attacks. EU also established the first EU-wide cybersecurity certification program to ensure that products, processes and services sold in EU countries complied with cybersecurity standards. The Act is the cornerstone for EU's cybersecurity protection. Besides, EU has also strengthened 5G security governance and issued a 5G network security risk assessment report to ensure highly secured 5G networks across EU.

7.5.5 *The United Kingdom*

Based on the existing laws and regulations, the British government has further improved its cyberspace governance system, refined management regulations, deepened cooperation with the industry, and strengthened the research and management of new technologies such as artificial intelligence and blockchain. The specific measures are as follows:

(1) Approving *Counter-Terrorism and Border Security Act 2019*, which increases penalties for terrorist acts and lists browsing terrorist content online, as illegal and criminal acts punishable by up to 15 years in prison.
(2) Issuing *Age Appropriate Design: A Code of Practice for Online Services*, which restricts social media companies from collecting, sharing and using children's personal data, and protects children with default measures such as "top privacy protection". Introducing *Ae Verification Regulations*, which stipulates that websites with pornographic materials must have age verification checks, otherwise they will face sanctions.
(3) Issuing *Online Harms White Book*, which states that the United Kingdom will build its independent regulatory agency to supervise social media, search engines, communication programs, and even file-sharing platforms with quantified penalties.
(4) Accelerating cybersecurity talent training and building cybersecurity center. Establishing national cybersecurity committee, investing £2.5 million to train professionals to strengthen British cybersecurity capability. Developing plans to build cybersecurity center to provide relevant information and intelligence analysis for the British military.
(5) Stepping up research on new technologies. Establishing the world's first Center for Data Ethics and Innovation (CDEI) to study inherent algorithmic bias in decision-making systems, and funding a group of companies on artificial intelligence and dark web.

7.5.6 *France*

France accelerates its participation in international cyberspace governance, and advocates its propositions on cyberspace governance through digital taxation, social media supervision, and *Paris Call for Trust and Security in Cyberspace*, in an attempt to obtain leading position in Internet governance in European Union. French president Macron released *Paris Call for Trust and Security in Cyberspace* at the 13th United Nations Internet Governance Forum (IGF) in November 2018, proposing to strengthen the prevention and response capabilities to malicious cyber activities, jointly combat infringement of intellectual property rights on Internet, prevent the proliferation of malware and technologies, ban online mercenary activities and attacks by non-state actors, and boost the development of international standards related to cyberspace security. France has further enhanced its own cyberspace governance, specifically in the following three aspects.

(1) Strengthening social media supervision. In January 2019, French government required social media to fulfill "duty of care", taking responsibility for and strengthening the review of platform content, to jointly manage pornography, terrorism and hatred speech on the platform.
(2) Controlling false news. French parliament passed *Anti-fake News Bill*, granting courts the power to demand the media should delete fake news during the election campaign to ensure that national elections were protected from misinformation. Those who violate the law will face one year in prison and a fine of €75,000.
(3) Levying digital tax. In July 2019, French president signed *Digital Tax Bill*, allowing the government to impose digital tax on Internet companies that realized the annual global revenue of over €750 million with over €25 million from France. The tax is 3% of the revenue from French market.

7.5.7 Germany

Germany has enhanced its Internet supervision and governance, imposed stricter regulatory requirements on social platforms and network operators, stepped up efforts to combat cyber crimes, and strengthened the protection of youth online. The specific measures are as follows:

(1) Strengthening management responsibilities of social platforms and network operators. In February 2019, Germany announced a restriction on the U.S. social media Facebook, forbidding it to collect user data from third-party services. Facebook renders its service on condition that the users allow it connect to and import data from third-party websites or mobile Apps.
(2) Enhancing tax management of Internet companies. In February 2019, German Federal Ministry of Finance proposed to impose a 15% withholding tax on online advertisements on foreign online platforms. This move will give German government the power to tax multinational Internet companies.
(3) Stepping up the crackdown on the dark web market. In March 2019, German Bundesrat took a vote and defined the provision of technological support for dark web platforms as criminal offense. Platform technology providers are required to monitor illegal activities such as trading drugs, explosives, and

disseminating child sexual abuse materials. Any violation is subject to criminal penalties of up to three years in prison.

(4) Improving laws and regulations on online youth protection. In June 2019, Germany passed an amendment to criminal law, aiming to protect young people from online sexual abduction. Any violation will constitute a crime and is subject to 3–5 years in prison.

7.5.8 *Russia*

Faced with unfavorable international landscape and harsh external environment, Russia attaches great importance to ensuring cybersecurity and cyber sovereignty, and achieves substantial progress. In March 2019, Russian president Vladimir Putin signed *Fake News' Law* and *Disrespect Laws*, stepping up efforts to crack down on words and actions that spread false information and disrespected the country. *Fake News' Law* prohibits individuals and legal persons from publishing and disseminating false information on Internet that may cause serious consequences, and stipulates that legal entities or individuals who repeatedly spread fake news that disrupts public order and security will face a fine of up to ₽1.5 million. *Disrespect Laws* stipulates that anyone who publicly posts online statements expressing "disrespect for state symbols of the Russian Federation, the Russian Constitution and national authorities" will be fined or imprisoned for up to 15 days. In April 2019, Russian State Duma approved an amendment to *Russian Federation Communications Law* and *Russian Federation Law on Information, Information Technology and Information Protection*, proposing to develop an autonomous Russian Internet, autonomous root servers, autonomous domain name resolution systems, autonomous routing nodes, autonomous overall coordination and management agencies, autonomous Internet disconnection, etc., The amendment is also known as *Sovereignty Internet Law*. The law means that Russia is using legislative means to improve its autonomy level and control of key network infrastructure, which may have a major impact on America's current dominance in the distribution of fundamental Internet resources.

7.5.9 *India*

India speeds up its digitalization. On the one hand, India has revised and improved the existing laws and regulations to meet the needs of current cyberspace development. On the other hand, it has drafted and formulated new laws to regulate cyberspace behavior and enhance cybersecurity defense capability. In June 2019, Indian Federal Cabinet approved *Aadhaar Amendment Bill 2019*, stipulating that Indian ID agency shall issue a string of 12-digit random numbers to citizens to be used in combination with their basic demographic and biometric data, which will constitute the world's largest biometric ID system. In addition, India amends *Information Technology Law*, which stipulates censorship of social platform to review content deemed inappropriate by the government, and imposes heavier penalties on online fake news and child pornography. Indian government also discusses *Banning of Virtual Currency Restrictions and Official Cryptocurrency Regulation Bill 2019*, which if passed, will deem possessing, buying and selling bitcoin and other digital currencies as crimes.

7.5.10 Brazil

Internet in Brazil develops rapidly. Investment in R&D and innovation in emerging technologies such as Internet of Things and artificial intelligence continues to rise, and Internet commercial applications become increasingly popular. In addition, Brazil pays more and more attention to Internet governance and personal information protection. *Lei Geral de Prote o de Dados Pessoais* (LGPD), promulgated in 2018 was expected to come into effect on February 15, 2020. This law integrates more than 40 relevant laws and regulations on different levels across the country, creating a new legal framework for the use of personal data. It not only protects personal rights, but also promotes the use of clear, transparent and comprehensive personal data rules to boost economic, technological and innovative development. In July 2018, Brazil established a digital governance and information security committee, responsible for monitoring data processing and supervising the implementation of strategic ICT plan. The committee has planned a national ICT development strategy, including increasing the use of information technology in public administration, encouraging public participation in decision-making and expanding access to public information. Brazil plans to establish a national data protection agency and list data protection as a basic right of citizens under the Constitution.

Throughout 50 years of development, Internet has become a growing force in driving human society from industrial civilization into information civilization. Faced with profound restructuring and changes in the current international order, international community needs to work together to promote mutual trust and open cooperation in international cyberspace governance, improve governance rules, establish effective cooperation mechanisms, and maintain the order of global cyberspace, to ensure that Internet continues to contribute to the development of human civilization for the next 50 years, and build a community with a shared future in cyberspace for the benefit of human.

Chapter 8
International Cyberspace Governance

8.1 Overview

At present, the international governance of cyberspace is a major realistic issue confronted by all countries and various actors. To meet the challenges from new technologies and new business and the need to build peaceful, open, secure and cooperative cyberspace, the international community is increasingly concerned with international cyberspace. In the past year, the complexity of that respect has gradually emerged, as shown in the following aspects: the formulation of norms for global cyberspace governance is relatively lagging behind, the differences in global cyberspace governance models still exist, and trust among actors needs to be strengthened.

8.2 Current Problems with International Cyberspace Governance

As international cyberspace governance is moving into the deep-water zone, there are an increasing number of difficulties and challenges. The norms of the governance are still incomplete, and the dispute over the concept of "multi-stakeholder model" and "multilateral model" has not yet come to an end, and the trust among various actors needs to be strengthened. Although the international community has strong expectations for the reform of the Internet governance mechanism, there are difficulties in practice. Cybersecurity and cyber abuse have not been effectively resolved. The potential risks brought about by the development of the digital economy and AI have given rise to new governance problems facing the international community.

Chinese Academy of Cyberspace Studies, *World Internet Development Report 2019*, https://doi.org/10.1007/978-981-33-6938-2_8

8.2.1 Immediate Need to Solve the Laggling-Behind Standardization of Global Cyberspace Governance

The relatively lagging-behind formulation of global cyberspace governance standards has restricted policy and capacity coordination for global cyberspace governance. Although more and more actors acknowledge that global cyberspace has become a new area of human activity, it is still difficult to reach a consensus on common norms with broad adaptability. It is an important issue facing the international community whether it is possible to formulate the standards and their implementation mechanism with a multilateral and multi-actor participation based on mutual trust and dialogues and collaboration.

8.2.2 Differences in Governance Models Needing to Be Resolved Through Negotiation and Coordination

The dispute over the model of global cyberspace governance is going on. How to form an effective consensus will produce great impact on the development of global cyberspace governance in the future. One of the most important factors affecting the formulation of international rules for global networks is the difference in dominant principles and models, that is, the difference between traditional "multilateral" and "multi-stakeholder" models. With the practice of international governance, more and more countries realize that the two models have no essential difference in terms of governance subjects, but in the status and role of the government in the governance structure. The multilateral governance model emphasizes that governments of various countries take the lead and that multiple stakeholders participate in the international cyberspace governance. The multi-stakeholder model emphasizes the diversity of governance subjects, including governments, businesses, technical communities, research institutions, social organizations, and individuals. These subjects will be treated equally in the governance. The two models actually indicate that different countries have different choices of governance paths according to their specific practices at different stages of the Internet development. The international community should negotiate or collaborate on this basis to promote multilateral/multi-stakeholder governance rather than block the building of the global governance system and capacity and hence the sustainable development of cyberspace due to the coexistence of the two models. Over the past year, with changes in the international environment of cyberspace, more and more people are aware of the rationality and possibility of the coexistence of the two models. Diversified subjects should be encouraged to strengthen dialogue and negotiation to promote shared governance and common prosperity in cyberspace.

8.2.3 *Trust Among the Actors to Be Strengthened*

Mutual trust can increase the depth and breadth of cyberspace governance cooperation and lead to mutual benefit and win-win results. However, with the asymmetric distribution of network information resources and different degrees of dependence on network technology, as well as different interests in cyberspace, different actors adopt different decision-making mechanisms in international governance. Lack of trust makes it difficult to cope with some global issues of cyberspace, such as cyber terrorism, cybercrime, and cyber abuse. It is necessary to strengthen mutual trust to build an effective international cooperation mechanism for cyberspace governance.

8.2.4 *Development of New Technologies and Applications Having New Demands for International Governance of Cyberspace*

Digital economy is increasingly becoming the focus of sustainable development in various economies around the world. There is yet no clear responsibility of the international governance mechanism and no perfect rules for the development of digital economy. More and more countries and regions have increasingly strong needs for data protection and network security, so they keep making legislation. Hence new policy barriers have been formed at certain levels. All these practices give rise to new challenges to the integration between digital economy and traditional industries, and how to eliminate the barriers is becoming a concern of the international community. New technologies and new applications, which are subversive and uncertain, urgently demand targeted governance tools and models for them.

8.3 Progress of International Cyberspace Governance Platforms

Platforms still play a role in global cyberspace governance affairs. Facing new situations and new issues in international governance in the past year, each platform has made positive efforts and achieved different results in accordance with its own capabilities and characteristics.

8.3.1 United Nations High-Level Panel on Digital Cooperation

Digital technology crosses national borders in a unique way, and cross-sector, cross-border cooperation is essential to tapping the full social and economic potential of digital technology and mitigating the risks that the technology may pose. As the world's largest international organization, the United Nations plays an irreplaceable role in the field of Internet governance. It established the High-level Panel on Digital Cooperation in July 2018 to strengthen cooperation in digital economy among governments, private sector, civil society, international organizations, and technological and academic circles. The newly launched high-level group will describe the development trend of digital technology, identify opportunities and challenges, and make recommendations for strengthening international cooperation in the digital field. It has set up offices in New York and Geneva, with 20 members from government, business, civil society and academia.

8.3.2 International Telecommunication Union (ITU)

The International Telecommunication Union (ITU) is the United Nations agency responsible for information and communication technology (ICT) affairs. In October 2017, ITU held the World Telecommunication Development Conference (WTDC-17), with the theme "Information and Communication Technology for the Sustainable Development Goals". The conference topics covered digital economy, network security, ICT technologies and applications, and the telecommunication market environment and supervision, statistics and other fields. In May 2018, the second Artificial Intelligence for Good Global Summit was held at the ITU headquarters in Geneva. The event focused on exploring the potential of AI to accelerate the realization of the United Nations Sustainable Development Goals (SDG). In November 2017, ITU released the ninth edition of the *Measuring the Information Society Report 2017*, ranking countries according to the level of their information and communication technology development.

8.3.3 World Summit on the Information Society

The World Summit on the Information Society (WSIS) is the most important platform for dialogue in the Internet field under the United Nations framework. It aims to explore how to make the digital revolution based on information and communication technology benefit the development of humankind. It is one of the world's largest community of information and communication technology development. In March 2018, WSIS Forum was held in Geneva of Switzerland, with 2,500 ICT experts from the world participating in it. It focused on the sustainable development of key SDG areas like health, starvation, education, youth, inclusiveness, gender equality, environment, infrastructure and innovation as well as their inclusive ICT measures.

8.3.4 Internet Governance Forum

The Internet Governance Forum (IGF) is the highest-level dialogue platform for global Internet governance discussion between governments under the United Nations framework. In December 2017, the 12th IGF Forum was held in Geneva of Switzerland, with the theme "Shaping Your Digital Future". The topics covered the future of digital governance, digital economy, the Internet and ICT and sustainable development goals, AI, big data, IoT, IPv6, false information, data protection, digital literacy, information acquisition, digital divide, etc. In July 2018, the IGF held its second public consultation and multi-stakeholder advisory group (MAG) meeting, at which reporting and discussions were made about global and regional Internet governance forum work, including proposal selection, special sessions, and best practice forums (covering AI, big data and IoT, cyber security, gender and access, local content, etc.), and the progress of the MAG Working Group.

8.3.5 Internet Corporation for Assigned Names and Numbers (ICANN)

The Internet Corporation for Assigned Names and Numbers (ICANN) is the coordinating body that maintains the Internet's unique identifier system and its secure and stable operation worldwide. Since its reform in 2016, ICANN remains one of the most important international organizations in the field of Internet governance. In 2018, community opinions were collected on GDPR deployment, Internationalized Domain Names (IDN), and Root Zone Key Signing Key (KSK) rotation. While enhancing its own capacity building, it also strengthens coordination with other institutions. For example, it signed an MOU with the Global System for Mobile Communications Association (GSMA).

8.3.6 World Internet Conference

The World Internet Conference is a worldwide Internet event initiated by China and held annually in Wuzhen, Zhejiang. In December 2017, the Fourth World Internet Conference was held with the theme "Developing Digital Economy for Openness and Shared Benefits: Building a Community of Shared Future in Cyberspace".

The conference focused on five aspects: digital economy, cutting-edge technology, Internet and society, cyberspace governance, and communication and cooperation. *China Internet Development Report 2017* and *World Internet Development Report 2017* were released for the first time at the conference. The reports summarized historical achievements of Internet development, analyzed the characteristics and the status quo of Internet development, and established and released the World Internet Development Index System and China Internet Development Index System. The conference also released two fruitful documents: *Wuzhen Outlook* and B*est Practice Cases of World Internet Development 2017*. In November 2018, the Fifth World Internet Conference was held, at which important results such as *China Internet Development Report 2018*, *World Internet Development Report 2018*, and *Wuzhen Outlook 2018* were released.

8.3.7 World Economic Forum

The World Economic Forum (WEF) is an unofficial international organization that studies and discusses problems in the world economy and promotes international economic cooperation and exchanges. According to the WEF's Global Risk Report 2018, cyberattack will be the third biggest risk worldwide in 2018, and cyber risks are affecting society and economy in new broader ways. In January 2018, at the annual meeting with the theme "making the common destiny in a differentiated world", WEF announced the formation of a new global network security center, to strengthen the existing measures (such as Cyber Resilience: Playbook for Public-Private Collaboration), to establish a library of best practices, to improve the partner's understanding of network security, to promote the supervision framework and to construct a future network security landscape think tank and thereby to build secure and reliable global cyberspace.

8.4 Participation of Traditional International Organizations in the International Cyberspace Governance

Over the past year, traditional international organizations have increasingly taken international governance of cyberspace as their important work content, which shows that the pattern of global governance is undergoing continuous and profound changes.

8.4.1 Group of Twenty Finance Ministers and Central Bank Governors (G20)

In 2017, G20 Summit held in Hamburg of Germany attached great importance to cyber security and issued a number of documents related to cybersecurity governance. In particular, participants in the summit proposed that G20 should play a key role in critical infrastructure protection related to global economic stability, the fight against cybercrime and cyberterrorism, and other fields. In the declaration of the 2017 finance ministers' and central bank governors' meeting, participating countries focused on transparency and security in the context of digital economy, and set specific goals on how to build a secure network environment, develop secure information infrastructure, and enhance trust among countries in the development of the digital economy. At the second G20 Consumer Summit held in Argentina in June 2018, the challenges and opportunities for consumers in 2018 and beyond were discussed, including new challenges and security of children's access to the Internet.

8.4.2 BRICS

In 2018, the 10th BRICS Summit issued the *Johannesburg Declaration*, in which the participating countries reaffirmed the responsibility of all countries for preventing terrorism, cyber financing and terrorist actions within their own territories. They emphasized that all countries should enhance international cooperation, and counter terrorism and crime with ICT and that they should formulate a universally accepted legal instrument to combat ICT crime within the framework of the United Nations, while recognizing the progress made in cooperation in accordance with the *BRICS Roadmap of Practical Cooperation on Ensuring Security in the Use of ICTs* or other consensus mechanisms.

8.4.3 Asia-Pacific Economic Cooperation (APEC)

In November 2017, the 25th APEC Leaders' Informal Meeting adopted the *APEC Internet and Digital Economy Roadmap* and the *APEC Cross-Border E-Commerce Facilitation Framework* to strengthen cooperation in Internet and digital economy, bridge the digital divide and share the digital dividend with more people. APEC has made significant progress in building a green supply chain cooperation network and an Asia-Pacific model port network, which has been unanimously recognized by member states.

8.4.4 Shanghai Cooperation Organization (SCO)

Cyber security with counterterrorism and international codes of conduct shaping as the core is always the concern the Shanghai Cooperation Organization (SCO). According to the relevant resolutions of the SCO Regional Counter-Terrorism Agency Council, the SCO Xiamen-2017 anti-cyber terrorism exercise was held in December 2017. The delegations of the competent bodies of the eight member states and the delegations of the executive committees of regional anti-terrorism organizations participated in the exercise, which helped to improve the coordination capabilities of member states in terms of cyber security information sharing and action. In January 2018, a routine meeting of the International Information Security Expert Group of SCO member states was held in Wuhan of Hubei. In June 2018, cybercrime combat was taken as one of the three major focuses at the Qingdao Summit of SCO. According to the declaration at the summit, SCO member states will strengthen cooperation in the light of new regional security situation and cyber security issues, and coordinate law enforcement agencies in intelligence information, judicial assistance and law enforcement standards.

Columnn 10: Non-state Actors

In recent years, non-state actors have been playing an active role in international cyberspace governance. An important reason is that they have realized that the integrity and improvement of the international cyberspace governance system is of significance to the interest of non-state actors. The participation of non-state actors in the international cyberspace governance is of long-term importance to the creation and improvement of the global governance environment. In April 2018, 34 technology and security companies, including Microsoft, Facebook, Cisco, Hewlett-Packard, Nokia, Oracle, Symantec, British Telecom, and Telefónica, signed a network security technology agreement, pledging to protect users' information and not to help governments launch cyber-attacks. In addition, Microsoft has released a guidebook on the *Cybersecurity Policy Framework*, suggesting that businesses and governments should conduct more cooperation in the development of security technology standards, information sharing, establishment of mechanisms and specifications.

In February 2017, the Global Commission on the Stability of Cyberspace (GCSC) officially presented itself at the Munich Security Conference, which attracted wide attention. At present, the committee is composed of more than 40 well-known cyberspace figures from nearly 20 countries. Its main sponsor is the Dutch government. Its secretariat is located at The Hague Centre for Strategic Studies (HCSS) in the Netherlands and East-West Center in the United States. On November 21, 2017, the first rule issued by the Commission was to defend the Internet's public core. It means that without affecting their own rights and obligations, states and non-state actors cannot do or condone intentional material damage to the generality or integrity of the Internet's public core and hence to the cyberspace stability. In addition to gaining support from the commission members, this rule has been recognized by civil society groups such as the Electronic Frontier Foundation (EFF) and Internet pioneers such as Vint Cerf. Therefore, it has produced great impact.

8.5 Internet Governance in Some Countries and Regions

8.5.1 The United States Taking the Lead in Top-Level Design of Cyberspace Strategies and Data Security Protection Legislation

The United States attaches great importance to the strategic value of data resources, having successively introduced national strategies, implemented supporting measures, and systematically promoted the development of national big data. By strengthening data security protection, the country is improving personal information and data legislation, protecting the overall network and national security, national data rights and its citizens' rights. In March 2018, U.S. President Trump officially signed the *Clarifying Lawful Overseas Use of Data Act* (also known as the "CLOUD Act"), which provides a legal basis for law enforcement agencies' acquisition of data stored on overseas servers, while allowing "eligible foreign governments" to issue orders on data collection to organizations within the United States directly after signing administrative agreements with the US government. In June 2018, the State of California introduced the *Consumer Privacy Act*, which is considered by far the most stringent data privacy bill in the United States.

In September 2018, the United States issued the *National Cyber Strategy*, which proposes promoting the multi-stakeholder model of Internet governance, building interoperable and reliable Internet infrastructure, and promoting and maintaining the country's innovation markets around the world.

In late 2017, the U.S. Federal Communications Commission (FCC) decided to remove net neutrality protections, sparking a global debate about platform and data neutrality. Although there is no net neutrality at the international level, many global operators are based in the United States, and the U.S. decision might affect the development of the global network and practices of many other countries.

8.5.2 China's Continued Pragmatic Participation in Cyberspace International Governance

In December 2017, President Xi Jinping wrote a letter to congratulate the opening of the Fourth World Internet Conference and emphasized that China advocates the "four principles" and "five proposals", and hopes to work with the international community to respect online sovereignty and promote the spirit of partnership. According to Xi, all matters should be tackled through consultation to ensure common development, mutual maintenance of security, co-governance, and shared results.

China is always ready to engage in bilateral/multilateral cyberspace cooperation at different levels in various fields. In May 2018, the China-Germany consultation on cyber security under the framework of the two countries' high-level security dialogue was held. During the consultation the two sides exchanged views on the situation of cybercrime, legislation on cybercrime and security, and cybercrime and cyber terrorism combating. In July of the same year, the fifth round of China-Germany governmental consultation was held, during which the two sides deepened exchanges and cooperation in terms of cyber security and in secure cross-border transmission of data. Significant progress has been made in the exchanges and cooperation in cyberspace governance since then. In June 2018, China and Russia signed the *Joint Declaration of the People's Republic of China and the Russian Federation*, according to which the two countries will expand exchanges in information and communication technology and digital economy, improve the interconnection of information and communication infrastructure, strengthen the cooperation in radio frequency and satellite orbit resource management, promote cyberspace development of two countries, and deepen mutual trust in cybersecurity. In September 2018, the Ninth China-EU Dialogue on Information Technology, Telecommunications and Informatization was held in Beijing. The two sides reviewed the progress of China-EU cooperation in the field of information and communication after the eighth dialogue, and had in-depth exchanges on ICT policy and digital economy, ICT supervision, 5G R&D, and industrial digitalization.

China values and supports multilateral governance at the regional level, and participates in and promotes the integration of cybersecurity and development of digital economy into regional cooperation. In June 2018, the Qingdao Declaration issued by SCO called on the United Nations to play a central role in formulating universally accepted international rules, principles, and norms for the responsible conduct of states in information space. In September 2018, the tenth conference of BRICS leaders was held and a consensus was reached on promoting practical cooperation in cyber security and combating terrorism and crime through the abuse of ICT.

8.5.3 EU Legislation to Strengthen Personal Data and Digital Copyright Protection

EU always attaches importance to cross-border data security policies and legislation. After GDPR went into effect in May 2018, EU immediately accused Google and Facebook of forcing users to agree to data collection, and would impose high fines. The strong law enforcement by EU has played a good warning and urging role, since Internet businesses such as Microsoft, Facebook, and Twitter all adjusted their privacy policies to improve their user data protection. GDPR has led to legislation on personal data in various countries, such as the United States, India, and Brazil.

EU is enhancing the governance of digital copyright. In September 2018, the European Parliament approved through voting the *Directive on Copyright in the Digital Single Market*, which proposes the amendment of digital copyright regulations and allows news media to charge technology platforms for their copyrighted materials to protect the interests of Internet content creators and to facilitate Internet businesses' protection of digital copyrights.

8.5.4 The United Kingdom Attaching Importance to Cyber Counter-Terrorism and Achieving Results in Cracking Down on Dark Websites

The United Kingdom is committed to strengthening cyber anti-terrorism governance. In June 2017, the British Prime Minister Theresa May called for the re-formulation of standards for social networks including Google, Facebook and Twitter to prevent terrorist activities in cyberspace. In September 2017, the United Kingdom, France and Italy decided to shorten the time limit for removal of Internet terrorism content to two hours.

The United Kingdom has established organizations jointly combating criminals in dark web. In 2017, National Crime Agency set up a Dark Web Intelligence Department and cooperated with other cyber intelligence and enforcement agencies in combating criminal activities on dark websites. In July 2017, over ten countries, including the United Kingdom, the United States and Canada, jointly closed down the world's largest dark web platform AlphaBay, which engaged in the trading of drugs, weapons and other illegal items. Thanks to the joint effort of multiple countries, AlphaBay, RAMP, Dream Market and Hansa Market, the top four dark trading markets, have been closed.

8.5.5 A Multi-level Internet Governance System Formed in Russia

Russia has formed a multi-level Internet information governance system with laws, institutions and technologies as the main body. In recent years, the country has successively issued a series of documents such as the *Concept of Russian Network Legislation, Legislation Development Concept in the Field of Information and Informatization of the Russian Federation*, and *Doctrine of Information Security*, and amended more than 20 laws including *National Policy Law of Development and Utilization of the Internet* and *Law of Right to Information.* As a country that strictly restricts the flow of cross-border data, Russia has successively promulgated the *Law of Information, Information Technology and Information Protection* and the *Personal Data Law of the Russian Federation*, and basic rules for local storage of data.

Russia is building cyberspace governance platforms. In July 2018, the International Cyber Security Conference hosted by the Savings Bank of the Russian Federation and co-organized by the Digital Economy Organization and the Association of Russian Banks (ARB) was held. The meeting focused on the organized cybercrime combat, cyber risks in the digital transformation era, and strategic assessment and protection of confidential information, blockchain technology and information security. Russian President Putin attended the meeting and called on all countries to establish a unified "rule of the game" and international standards in the digital technology industry.[1]

8.5.6 Rapid Internet Development and Data Protection Legislation in India

India's Internet and communications industry has developed rapidly. In 2015, the number of Indian Internet users rose to more than 400 million, higher than that of the United States, making it the second largest Internet country after China. In November 2017, the Fifth International Conference on Cyberspace was held in India and Prime Minister Modi was present. He emphasized the importance of cybersecurity and personal privacy protection. In July 2018, shortly after EU implemented GDPR, India promulgated the *Personal Data Protection Law* (Draft). In the same month, the coutnry's government approved the principle of net neutrality, claiming that any deviation from and violation of the principle of net neutrality will be severely punished. In 2018, the country participated in the SCO Summit for the first time as a full member. It will participate in the fight against cybercrime.

[1]Speech by the Russian President Putin at the International Cyber Security Conference. http://www.kremlin.ru/events/president/news/57957。.

8.5.7 *Cooperation in Data Privacy Protection Between Japan and EU*

According to GDPR, when EU member states transmit personal data to a third country, the latter must pass the EU's "adequacy" protection standard. In July 2018, Japan and EU concluded negotiations on the basis of mutual "Adequacy Decision", and the two sides agreed to start a procedure to identify each other as a region where "the protection of personal information is at the same level" and to ensure the security of data flow to promote the establishment of the largest global cross-border data flow zone. In September, the European Commission launched an "Adequacy Decision" procedure in Brussels, affirming Japan's additional guarantee measures applicable to EU personal data transferred to Japan and the Japanese authorities' commitment to obtaining personal data for law enforcement and national security. Japan promised that its data protection would be at the same level of that in EU. At the same time, Japan adopted a similar process to recognize the EU's data protection framework. This means that for that country, personal data is allowed to be transferred between EU and Japan without any specific authorization; for EU, personal data transmitted to Japan enjoys strong protection in accordance with EU privacy standards. Japan became the first Asian country to win the "Adequacy Decision" from EU.

Faced with the changes in the international cyberspace governance over the past year, countries, international organizations and other cyberspace governance bodies have become increasingly aware of the importance and necessity of building a future-oriented international cyberspace governance system based on mutual trust and co-governance. More and more forces in the international community have shifted from rational thinking to governance practices, from technological innovation to institutional innovation, and reached a consensus on building a peaceful, secure, open and cooperative international governance environment for cyberspace.

Looking ahead, the international community needs to further enhance mutual trust, communication and cooperation, and form a new pattern of co-governance. In order to achieve the UN 2030 Sustainable Development Goals, we should work together to build a community of shared future in cyberspace and create a better future.

Postscript

Throughout 50 years of remarkable development, World Internet has transformed the production and lifestyle of human, and opened new age in human history. The world today increasingly becomes a community of shared future, and our fates grow more intertwined every day. The sharing and co-governance of Internet around the world is not only an opportunity presented by history but also a mission of our time. We hope that *World Internet Development Report 2019* (hereinafter referred to as "The Report") has given a full account of world Internet development in the past year, interpreted the development trend from China's perspective, and provided Chinese solutions to the problems in this field. The Report intends to contribute to the joint effort on the development, security and governance of cyberspace, and promote the sharing of fruitful results thereon.

During the compilation of The Report, we received guidance and support from the Office of the Central Cyberspace Affairs Commission. Leaders of the Office of the Central Cyberspace Affairs Commission offered their insights, and all the affiliated agencies and units provided strong support for the preparation of The Report, especially with the provision of relevant data and material content. The Report is led and organized by Chinese Academy of Cyberspace Studies (CACS), and co-edited by National Computer Network and Information Security Administrative Center, China Academy of Information and Communications Technology, China Industrial Control System Cyber Emergency Response Team, Tsinghua University, Peking University, Beijing University of Posts and Telecommunications, Xi'an Jiaotong-Liverpool University, and other agencies. Main Contributors are: Yang Shuzhen, Fang Xinxin, Hou Yunhao, Li Yuxiao, Li Changxi, Liu Shaowen, Feng Mingliang, Chao Baodong, Li Zhigao, Tian Yougui, Long Ningli, Tang Lei, Li Min, Liu Yan, Jiang Wei, Nan Ting, Zhao Yanwei, Han Yunjie, Dong Zhongbo, Wang Hailong, Li Bowen, Shen Yu, Li Xiaojiao, Wang Meng, Wang Xiaoshuai, Ma Teng, Zhao Gaohua, Xie Yi, Li Wei, Xu Xiu'an, He Bo, Jia Shuowei, Yang Xiaohan, Sun Luman, Tian Yuan, Yang Shuhang, Xiao Zheng, Song Shouyou, Wu Wei, Zhang Qiyuan, Gao Ke, Chen Jing, Yuan Xin, Xu Yanfei, Xu Yu, Li Yangchun, Deng Jueshuang, Cai Yang, Wang Zhongru, Wang Hualei, Wang Liying, Qian Yiqin, Ding Li, Xu Yuan, Wang Xiaoqun, Wang Shiwen, Zhou Yu, Lou Shuyi, Meng Nan, Zhou Yang, Chen Kai,

Chinese Academy of Cyberspace Studies, *World Internet Development Report 2019*, https://doi.org/10.1007/978-981-33-6938-2

Mou Chunbo, Zhao Li, Jin Zhong, Wu Yanjun, Chong Dandan, Liu Shaohua, Li Shuai, Xin Yongfei, He Wei, Sun Ke, Zheng Anqi, Wang Mingzhu, Meng Qingguo, Xu Ji, Hu Shiyang, Li Yan, Xie Yongjiang, Liu Yue, Dai Lina, Fang Yu, etc.

The Report, though successfully published thanks to the strong support and considerable help from all sectors of society, is inadequate in terms of perspective and insight due to limited research experience and tight deadline. Therefore, we welcome opinions and advice from government departments, international organizations, research institutes, Internet companies, and social organizations across different sectors, home and abroad, to help us produce better Report in the future and contribute more wisdom and strength to world Internet development.

Chinese Academy of Cyberspace Studies (CACS)
September 2019

The manufacturer's authorised representative in the EU is Springer Nature Customer Service Centre GmbH, Europaplatz 3, 69115 Heidelberg, Germany. If you have any concerns regarding our products, please contact ProductSafety@springernature.com

Printed and bound by CPI Group (UK) Ltd, Croydon, CR0 4YY

15/07/2026

02167631-0002